# Railway People

George Turner Smith

Copyright © 2017 George Turner Smith

The moral right of the author has been asserted.

Apart from any fair dealing for the purposes of research or private study, or criticism or review, as permitted under the Copyright, Designs and Patents Act 1988, this publication may only be reproduced, stored or transmitted, in any form or by any means, with the prior permission in writing of the publishers, or in the case of reprographic reproduction in accordance with the terms of licences issued by the Copyright Licensing Agency. Enquiries concerning reproduction outside those terms should be sent to the publishers.

Matador
9 Priory Business Park,
Wistow Road, Kibworth Beauchamp,
Leicestershire. LE8 0RX
Tel: 0116 279 2299
Email: books@troubador.co.uk
Web: www.troubador.co.uk/matador
Twitter: @matadorbooks

ISBN 978 1785898 808

British Library Cataloguing in Publication Data.
A catalogue record for this book is available from the British Library.

Printed and bound in the UK by TJ International, Padstow, Cornwall
Typeset in 11pt Palatino by Troubador Publishing Ltd, Leicester, UK

Matador is an imprint of Troubador Publishing Ltd

*To Dad and Grandad Smith – railwaymen*

# Forward to 'Railway People'

A few years ago I had the idea of gathering together my various railway articles into one book. My thought was that arranging a compilation of what was already written would gain me an insight into the wacky world of publishing. As I had never written a book before, I had no idea how the process worked. Not surprisingly, I couldn't interest a publisher in the concept and decided to self-publish. This, however, offered a different benefit: I could now find out how publishing worked from the inside. Nevertheless, I knew I would have to keep the overall cost within my limited budget.

Given that most of the content was already written, I mistakenly believed that putting the book together would be both easy and straightforward. It was neither. For a start, the articles I chose to include had been written over a ten year period during which I effectively learned how to write. Consequently, the earliest articles were both poorly written and sometimes out of date. This meant that several were rejected immediately whilst others required substantial revision before they could be considered good enough for inclusion. By losing a great chunk of the material I had accumulated, I found that what remained was insufficient to make a book of reasonable length and, in consequence, I had to write additional chapters written from scratch, whilst seriously revising the remainder.

I have never been the most patient of people and revising old material bored me to tears. When I finished therefore, the clever thing to have done would have been to hand the book over to a third party for edit and proof-reading. This would, however, have required additional money which I didn't have so I let it go as it was. The consequence of this was that errors in presentation of the work prevented potential railway book reviewers taking it seriously. Nevertheless, there was still a lot of material remaining which I thought the general public would find interesting and the book duly appeared. But, since that day, at the back of my mind, I always had the intention, if I could ever raise the enthusiasm and find the time, to look at the book again. This would also afford me the opportunity to edit the book properly and perhaps add new material from magazine articles unpublished at the time the first book appeared. Since publishing that book, titled 'Those Railway People', I have written three more and believe I have now got my eye in. I am therefore ready to have another go at my first.

To those generous souls who purchased 'Those Railway People', I hope you like what I have done with it. What to leave out and what to put in caused me much grief. I decided, after much heart searching, to retain the piece about Dr. Beeching although, in fact, it didn't really fit because it was never a magazine article, but I axed a couple of other stories, which I thought repeated or overlapped with material already included. The additional chapters are all made up from articles that were originally published in 'Backtrack' and my thanks therefore go to Michael Blakemore, the editor, for all his support throughout the years.

George Smith 2016

# Introduction

I once mentioned to a friend I was thinking of writing a railway book and she said, 'Oh – a nerd's book'. I knew exactly what she meant. In my experience there were two types of railway book. The first, an engineering manual, consisting of page after page of technical detail – often with the original engineering drawings reproduced in agonising detail. For those readers desperate to know the precise wing mirror specification for a Gresley UB40 this kind of book is essential but, to the rest of us, such information is of limited interest at best and at worst gives you blinding headaches. The second type of railway book is designed for past and present train spotters, particularly those with a limited attention span – and I have to include myself in this category. Each page of this book is made up of pictures of locomotives, with the images separated by a minimal text. What written content is included is provided by authors of the first type of book but, by virtue of its limited nature, only induces mild pain behind the eyes. Unfortunately, after looking at the pictures, there is no reason to open these books again.

There are of course railway books out there which are both intelligent *and* readable – the biography of George Stephenson by Hunter Davies and most of those written by Tom Rolt spring to mind – but I would suggest that they are desperately few and far between. I find this sad because I love railways and (especially) steam locomotives and want everyone to see them in the same way I do.

I should point out, in fairness, I could no more write 'Book Type 1' than drive the 'Flying Scotsman' nor would my own locomotive pictures ever be good enough to grace the pages of 'Book Type 2'; I have, for information, a varied collection of loco images, where the object of the picture is either too far away to be in focus, already past the camera, too dark, too light or too boring – so perhaps there is a modicum of professional jealousy at work here.

Anyway, about the present book, I must confess there is far too much detail about obscure railways in the North East and much less about better known lines in other parts of the country. In defence, I can only say that, since I am from the North East, it is inevitable that I write mainly, if not exclusively, about places and events of which I have a little personal knowledge.

My background, for those of you who haven't read my Hartlepool book, is that I was brought up on a sprawling council estate in that town. My father worked on the docks as a railway carpenter and my grandfather was a railway porter. To compliment these genuine railway connections, I also discovered there was a Ralph Sotheran who chaired the first meetings of the wonderfully named 'Great North of England, Clarence and Hartlepool Railway' in the 1840s. Since my grandmother's name is Sotheran and her family lived in the old town for as long as records exist, I like to think, without any proof whatsoever, that I am somehow related. For these tenuous reasons I contend that railway blood runs in my veins.

Like most kids on my estate, from the age of five, I had two hobbies – football and train spotting. Being both long and short sighted, as well as asthmatic, I was never going to be another Bobby Charlton, so railways became my first love. Sadly, at the very time I started train spotting, steam was entering its terminal days. I therefore made it my personal mission to record, in crappy school exercise books, as many of the steaming, hissing, clanking machines I saw passing through the town before they disappeared for good – even when this meant skipping school to do so. My weekends were taken up standing below the railway arch at the end of my road, along with a gang of similarly enthusiastic urchins, waiting for the heart-stopping noise of an engine whistle. It was a big moment for all of us when the main line signal went up. In Hartlepool there were just two express trains a day. These were destined for London and were headed by top-link engines from York or Gateshead. To anyone who wasn't there, I can't possibly convey the thrill of expectation when we heard an engine approaching, especially if it was accompanied by the unique falsetto harmony of a double chimney (apologies to non-railway buffs here) or the distinctive haunting whistle or 'chime' which indicated an approaching streamlined 'streak' ( Gresley A4). Most often, however, the express was pulled by a less impressive loco from Gateshead engine shed whereupon we would shout 'Scrap it'.

Unfortunately the powers-that-be took us at our word.

Still that was then and now we can visit Heritage Railways, such as the Watercress Line and Severn Valley Railway, where you get the impression of what it was like to be there all those years ago – but without the heart-stopping thrills, since the express engines now plying these backwaters stroll along at a pedestrian 25mph to comply with century old legislation.

So why do I think this book is different to book types 1 and 2?

Well, primarily it is a book about the lives of people and not the intricacies and idiosyncrasies of machines; people with the same strengths and weaknesses as the rest of us. It was, after all, people like Dad and Granda, who made railways what they are (or were) and railway people are interesting precisely because they made the same mistakes and exhibited the same weaknesses and human frailties as you and I. In the early days, railwaymen mostly played it by ear, since there was no template for anything they did. They consequently learned by trial and error and often made mistakes. Yet, with each faltering step there was some improvement, however minor, and the railways moved inexorably onwards and upwards.

Railway people came from a variety of backgrounds and walks of life. In the following pages, for example, there are appearances by: George Stephenson, the semi-literate pit boy genius; John Rennie, the educated surveyor and engineer; William Huskisson, the Tory MP; Thomas Richardson, a former blacksmith, and so on. In the history of railways, all these people had a role to play and shared a common vision.

I salute them all and hope I have done them justice.

In two hundred years of railways, there is enough humour, pathos, drama and adventure to fill a million books and spawn a dozen Hollywood blockbusters. My earnest wish is that the reader of this book gets at least a glimpse of what's out there waiting to be discovered. My bottom-line is that I hope that the stories that follow are a 'good read' regardless of the reader's personal interest in railways and, consequently, not just another 'nerds' book of the type my friend referred to. If I failed, well at least I tried.

I would like to extend a big thank you to the editors of the magazines in which most of what follows appeared. Thanks also go to the staff of The National Railway Museum at York, the National Archive at Kew and the various local history societies, libraries and museums around the country who invariably went out of their way to help me. I also add thanks to my ever-patient wife, Maggie, who provided the best of the pictures, including the cover image. My grateful thanks also to my good friend Tom Horsburgh, who knows far more about railways than I ever could and to Jane Hackworth-Young who provided much of the material related to the Hackworth anecdotes. Finally, I extend my appreciation to all the railway people who brought such joy to my life.

# Contents

1. Branwell Brontë – station master — 1
2. Sir John Rennie – railway engineer — 12
3. Joseph Samuda – vacuum salesman — 24
4. Holman Stephens and the Selsey Tram — 35
5. George Stephenson and the Hartlepool Railway — 45
6. William Huskisson and the Rocket — 56
7. Robert Stephenson and the standing stones of Stanhope — 67
8. Thomas Richardson – steam to nuclear power — 81
9. George Graham's journal – a catalogue of errors — 90
10. Samuel Sidney's rural rides — 104
11. Christopher Tennant, railway entrepreneur — 122
12. Aled Roberts' life in railways — 132
13. Isombard Brunel and the 'Battle of Mickleton Tunnel' — 140
14. Dr. Beeching's railway reshape — 151
15. Nigel Gresley's soccer specials — 163
16. William Hedley's Wylam Dillys — 175
17. Timothy Hackworth's Archive — 190
18. John Hackworth's 'Tsar trek' — 201

| | |
|---|---:|
| 19. William Churchman and the 1911 Rail Strike | 213 |
| 20. J.R.Whitbread and the Railway Police | 222 |
| | |
| Postscript | 234 |
| Bibliography | 236 |
| Acknowledgements | 239 |

# Chapter 1

*Branwell Brontë – station master*

*Statue reputedly to be Branwell Brontë on the site of the former Luddenden Foot station*

On a grass verge on the edge of a small industrial estate in West Yorkshire stands a strange statue. The statue is an angular and somewhat stylised representation of a short Victorian man with long mutton chop side whiskers. He wears a long great coat and in his hand is a stone tablet inscribed with the words:

*'I cannot think as roses blow.'*

The whiskered gentleman stands on the site of the former Luddenden Foot station and the person represented in stone was once the Station Master, a certain Patrick Branwell Brontë, brother to Charlotte, Emily and Anne, the literary sisters of Haworth Parsonage. Branwell, as he was known to his family, like his sisters, aspired to be a writer and artist, so how did he end up organising the departure and arrival of trains at Luddenden Foot?

It was the dawn of the railway age. The Keighley and Worth Valley Railway, that would one day connect the lonely hill top village of Haworth to the rest of the world, had yet to be built, but in the Calder valley to the north, a more significant railway line was nearing completion. The aim was to link the expanding cities of Manchester and Leeds via a scenic railway through Hebden Bridge and Rochdale. A Parliamentary Bill authorising the railway acquired consent in 1824 but, because of fierce competition from the adjacent Calder, Hebble and Rochdale Canal Company, the authorisation wasn't enacted until 1837. George Stephenson was appointed Principal Engineer and, despite the severe nature of the terrain over which the railway was being built, construction work progressed smoothly until, in December 1840, completion of the line stalled because of structural failures in a three kilometre tunnel through the mountain at the line's highest point near Littleborough. Notwithstanding this hold-up, by the beginning of that year, the railway had reached Sowerby Bridge, one of the principal stations on the route. At this point, the proprietors, the Manchester and Leeds Railway Company, began advertising for staff for the stations along the way and, to the astonishment of friends and family, Branwell applied[1].

Patrick Branwell Brontë was Reverend Patrick's only son in a family of six. Although the fourth child, and a year younger than sister Charlotte, he was nevertheless viewed, as all sons were, as the heir to the family fortunes. Naturally intelligent and imaginative, he seemed destined to live up to family expectations. It was Branwell, for example, who led the games the children played and (with some of Charlotte's input) was the inspiration behind the imaginary world (Angria) around which the youngsters wove their early stories and on which they honed their writing skills. Nevertheless, if intelligent, he was also spoilt. He was, for instance, the only one of Patrick's six children not sent away to boarding school, but instead was kept at home and generally indulged by his father and doting aunt Elizabeth Branwell after whom he had been named. By the time he entered his teens, there was every reason to believe he would don the mantle of greatness his family expected. His downfall, when it came, had its origins in the tragic death, from consumption, of his older sisters, Maria and Elizabeth, to whom he was devoted. Without their presence and steadying influence, his personality slowly changed.

Branwell always had an inflated opinion of his own ability. His sense of 'self' accelerated after he was invited to join the Freemasons at the local lodge. Monthly meetings were held at the Black Bull in Haworth, just a few steps from the parsonage and it was there Branwell discovered the joys of intoxication, which must have been problematic for the local temperance society, of whom he was the secretary. Under the beguiling influence of alcohol, Branwell's natural inhibitions and shyness disappeared and 'Young Patrick', as he was known to his acolytes, became for a time the raconteur, intellectual and artistic bon viveur he believed himself to be. Unlike his introverted siblings, he also became a popular and respected member of this small community and seemed destined for a great future – if only he could find something he was both mentally and physically equipped to do. Given his literary precociousness, writing was the obvious choice but nobody seemed interested in any of the flowery poetry he wrote[2]. Still, he also had a gift for painting so his father paid for lessons in portraiture at Leeds, with the result that some of Branwell's paintings adorn the walls of the Parsonage today[3]. Sadly he blew away any future he might have had in portrait painting when he failed to attend an interview, arranged for him by his tutor, at the Royal Academy in London, having unwisely gone on a bender the night before.

Branwell was always hopeless with money. His failure to make a living as an artist meant borrowing from friends and acquaintances to further his drinking habit. By the beginning of 1840 he was both unemployed and broke and it was one of his drinking cronies, John Woolven, who suggested he apply for a job on the Manchester and Leeds Railway, where Woolven worked. Using Reverend Brontë and the landlord of the Black Bull as guarantors, Branwell consequently wrote to the Railway Company offering his services[4]. His application accepted, he was appointed Assistant Clerk in Charge at Sowerby Bridge station on the 31st August 1840 on the princely salary of £75/annum. Bearing in mind Branwell's artistic leanings, biographers down the years have expressed amazement at this unexpected career choice, as indeed did his family. Yet, perhaps to a hopeless romantic like Branwell, the attraction of railways was as obvious then as it was to people such as myself at the same age.

Sowerby Bridge, when Branwell went to work there, was not the quiet mid-Pennine backwater it is now but a significant centre for the wool industry. It sat astride a natural gateway for traffic between Lancashire and Yorkshire, where the river, an important canal and the railway all converged together, confined by surrounding hills[5]. When Branwell joined the railway payroll Sowerby Bridge was the existing terminus for trains from Leeds (passengers for Manchester being conveyed onwards by horse-drawn coach). The completion of the railway, via Stephenson's kilometre length railway tunnel through Blackstone Edge, did not take place until December the same year. Construction of Sowerby Bridge station was also far from complete and Branwell was forced to take lodgings in a local ale house called the Pear Tree Inn.

In accordance with the importance of Sowerby Bridge to the railway, the new station was not the windswept platform and plastic 'bus stop' shelters you see today but a grand U-shaped affair, designed to act as the railway's administrative centre, which could accommodate up to 800 railway employees[6]. There were many pubs in Sowerby Bridge and Branwell made a determined effort to get to know them all. The town was also only a couple of miles from Halifax, albeit on the other side of a steep hill, and it was there that Branwell's 'artistic' friends and, in his eyes, intellectual equals hung out.

Fortunately, since there were only four trains a day in each direction, Branwell had plenty of spare time to spend in the convivial company of friends and cronies. He took up writing again, using whatever company headed notepaper came to hand and it was while he was working there he finally became a published author, after a selection of his poems were reproduced in the Halifax Guardian; ironically the first of the talented Brontë children to be paid for their writing.

It may be surmised from these clandestine literary pursuits that, perhaps, Branwell's mind wasn't fully committed to railway work yet he must have coped with his duties adequately enough because, a few months after joining the railway, in April 1841, he was appointed Senior Clerk at Luddenden Foot, the next station up the line in the direction of Manchester.

*Northern Rail 'metro' DMU bound for Manchester passes site of Luddenden Foot station*

For 'Senior Clerk' read 'Station Master' and Branwell's salary doubled overnight to £130/annum. He was even given an assistant, John Walton, to help him out since he was now responsible for more trains since the line was doubled west of Luddenden station. The family (and Charlotte in particular, as is notable from her correspondence at the time) was pleased to see him 'getting on' in this unexpected career. Despite the extra work and responsibility, his job was hardly over-taxing and left him lots of time to indulge his two great interests, writing and drinking. In short he ought to have been content but, being Branwell, he wasn't. Winifred Gerin, in her biography of Branwell, refers to the 'cavernous chill' and 'towering heights' of Luddenden

Foot but in reality it is a pleasant enough village, little different to most small mill towns in the Pennines and, irrespective of Branwell's complaints about loneliness and isolation, it was less than a mile from his former friends and haunts in Sowerby Bridge. Luddenden station no longer exists but was located on a tight bend in the line, on an excavated platform cut into a steep bank, just a few feet above the nearby fast-flowing River Calder. There was again no accommodation ready for the station master so Branwell, took lodgings in Luddenden village a mile or so up the hill from the station.

As Senior Clerk he was responsible for the day-to-day running of the station and, worryingly, also in control of station finances, duties he initially seems to have performed to the company's satisfaction[7]. Despite appearance it looks, from Branwell's diaries and correspondence, that he was bored witless. The state of his mind at Luddenden may be gathered from a poem he wrote at the time, as he gazed from his office window which overlooked the line:

> 'Beneath the thundering rattle shook
> of engines passing by;
> The bustle of the approaching train
> Was all I hoped to rouse the brain
> of stealthy apathy'

Hardly a ringing endorsement of his career choice.

In the display units at The Brontë museum at Haworth there are other examples of the poems he composed at Luddenden Foot, all written on his employer's headed notepaper. His official station account ledger was similarly bedecked with his poems and drawings; including a full page caricature of John Murgatroyd, the local mill owner and one of his regular boozing buddies.

At the end of each working day, Branwell slogged up the hill to Luddenden where, in the lounge of the Lord Nelson public house, he was able to again don the mantle of an overlooked artistic genius[8]. Not surprisingly, as a regular feature of the local taverns, he was well thought of in the village and, being formally educated, found his services called upon to write letters on behalf of his less literate drinking chums. One of the better educated of these was a fellow railway employee called Francis Grundy. It was through Francis that Branwell was introduced to George Robert Stephenson, son of Robert Stephenson, whose more famous brother, as mentioned above, was the Manchester and Leeds Railway engineer. Francis, it seems, was singularly impressed with the intellect of the Brontë son, and advised him he was just wasted in his current employment, words which no doubt pandered to Branwell's own jaundiced view of his status. In Francis's words Branwell was:

*'alone in the wilds of Yorkshire with few books, little to do, no prospects and wretched pay.'*

Sentiments with which Branwell heartily concurred.

It wasn't long before the Brontë family favourite began to delegate such work as remained, in the afternoon, to his assistant Walton so he could get an early start at the Lord Nelson[9]. This was Branwell's undoing. Left to his own devices Walton dipped his hands in the till, and, when the company annual audits took place in January of 1842, the accounts were deficient to the tune of £11-1s-7d.

Branwell was summoned before the Board to provide an explanation but had little to offer. His ledger was produced and the Board were unimpressed with the drawings and poems that occupied the space where pounds, shillings and pence should have been. His interrogators, with apparently no appreciation of great literature, dismissed him along with the more culpable Walton and the missing monies subsequently deducted from Branwell's final salary. No second chance was offered; what seems today like heavy-handed treatment for a first offence was by no means unusual at the time. If the committee reports of the M&LR are to be believed, hiring and firing was a daily occurrence, one regular cause of dismissal being:

*'inaccuracy and carelessness in the discharge of (their) duties'.*

Still, it came as a blow to Branwell, not to mention the landlords of the many local hostelries he frequented. The pub owners got together, with a few of the more upstanding citizens amongst Branwell's drinking chums, and composed a 'round robin' to the M&LR demanding his re-instatement. The submission was duly considered and summarily rejected and Branwell found himself once again unemployed and penniless. Unfortunately, he was also now a drug addict. He had started taking laudanum (opium solution supplied by chemists as a pain killer) during his time at Luddenden as an 'economy measure'; laudanum being cheaper and more effective than gin in inducing oblivion. Sadly, drug addiction was to be Branwell's only legacy from his short time as a railwayman.

He didn't look back on his time at Luddenden Foot with any fondness:

*'I would rather give my hand than undergo again the grovelling carelessness, the malignant yet cold debauchery, the determination to find how far mind could carry body without both being chucked into hell, which too often marked my conduct when there.'*

Hardly the model employee.

Having returned home in disgrace, he took whatever work was on offer, including a teaching job from which he was rapidly dismissed after unwisely having an affair with his employer's wife. His options dwindling and his time with the L&MR the only full time employment on his CV, in desperation he applied to the Manchester and Hebden Bridge, Keighley and Carlisle Junction Railway, for the post of secretary, conspicuously name-dropping his casual Halifax acquaintance, George Stephenson, in the hope that even a tangential association with such a famous name would outweigh his unspectacular employment record. It didn't.

The Brontë family connection with early railways didn't quite end there. Not long after Branwell returned to the Haworth Parsonage his aunt Elizabeth died bequeathing her shares in the York and North Midland Railway (Y&NM), worth £900, to the three surviving Brontë sisters, conspicuously ignoring Branwell[10]. The Y&NM was the railway on which George Hudson built his reputation but, even in 1846, the star of the Railway King from York was on the wane. The control and management of the shares was entrusted to Emily but Charlotte could see that the railway bubble was about to burst and urged her younger sister to sell the shares and invest the money in:

*'some safer, if for the present less profitable, investment.'*

However, no move was made to sell them and, as Charlotte predicted, their value sank without trace along with George Hudson. The money the sisters hoped to gain from the sale had been earmarked to fund the purchase of a private school in Haworth, in which all the sisters (and perhaps Branwell) were meant to teach. The plan never reached fruition for reasons unrelated to the fall in value of the shares, namely a shortage of students willing to brave the climb to the wuthering heights of Haworth.

The 1841 census, completed when Branwell was living at Luddenden, shows the Reverend Patrick, his sister Elizabeth and the three girls Charlotte, Emily and Anne all resident at the Haworth parsonage. By the 1851 census only the old man and Charlotte remained. In one terrible eight month period, from Autumn 1848 to Spring 1849, first Branwell, then Emily, then Ann died; the girls from consumption and Branwell from debility (and probably consumption) aggravated by his opium addiction. The previous two years had been rewarding times for the sisters as they had four books accepted for publication between them[11]. Their brother, by then, was oblivious to their achievements.

*Haworth station, Worth Valley Railway*

Indeed, up until the day he died, at the tragically young age of 31, he was unaware that his sisters had achieved the greatness he always believed was his by right. Charlotte lived a further 6 years but was still only 39 when she too passed away. Her father, the last surviving member of the family, died in June 1861.

Three months after Reverend Patrick's death a meeting was held in the room over Branwell's old haunt the Black Bull; the very room in fact where the local freemasons had once welcomed Branwell into their fold. The outcome of the meeting was the creation of the Keighley and Worth Valley Railway. The outside world was finally coming to Haworth.

It arrived too late for Branwell. If he had lived he would still only have been in his early forties when trains began passing through the little railway station in the valley below the Parsonage which would have needed a Station Master. There is little point in speculating as to whether or not Branwell fitted the bill. To get to the railway station Branwell would have had to pass both the Black Bull and the dispensing chemist who supplied his laudanum so perhaps it wouldn't have worked out. Of the talented Bronte family, only Charlotte received the acclaim in her lifetime, if not the fortune, from the novels the Brontë sisters wrote. The legacy left by Branwell, former station master of Luddenden Foot, who promised so much as a child, was a peculiar stone statue next to the railway line in a unassuming Yorkshire mill village[12].

**Notes to Chapter 1**

1. Charlotte wrote to her friend Ellen Nussey with more than a hint of embarrassment, 'A distant relation of mine (sic) one Patrick Boanerges has set off to seek his fortune in the wild, wandering, adventurous, romantic, knight errant like capacity of clerk on the Leeds and Manchester Railroad.'
2. Wordsworth, one of Branwell's heroes, declined to even acknowledge receipt of the poems sent to him for comment.
3. The famous portrait of the three Bronte sisters kept at the National Portrait Gallery was painted by Branwell. The shadowy figure barely visible in the gap between Charlotte and her siblings is thought to have been a self-portrait of the artist which, for reasons unknown, he subsequently painted out.
4. The Manchester and Leeds however rejected the surety of Thomas Sugden, landlord at the Black Bull, and Branwell then had to rely on his aunt and namesake (Elizabeth Branwell) to come to his rescue.
5. The Calder, Hebble and Rochdale Canal.
6. Only the former ticket office and waiting room survives today, set back from the current station and now a licensed café (the 'Jubilee Restaurant'), the rest was destroyed in a mysterious fire that occurred, as they always seem to, just days before the old station was about to become a listed building.
7. A letter from Branwell to the M&L Railway committee is referred to in their minutes, wherein Branwell suggests a number of improvements that could be made to the Luddenden Foot station. His suggestions were approved.
8. There was also an extensive library at the Lord Nelson where he devoured the 'classics'.
9. It is perhaps no coincidence that one of the poems Branwell was working on was an ode to Lord Nelson – the man not the pub.
10. Branwell was conspicuously ignored.
11. Anne ('Agnes Grey' and 'The Tenant of Wildfell Hall'), Emily ('Wuthering Heights') and Charlotte ('Jane Eyre').
12. The statue, created by Mike Williams, was placed there as one of 11 waymarks for the Calder Way footpath.

# Chapter 2

## *Sir John Rennie – railway engineer*

*Portrait of Sir John Rennie*

If you were to stop a complete stranger and ask if he or she had ever heard of George Stephenson even the least knowledgeable would probably know that the man had something to do with railways. Ask the same question about Sir John Rennie and you would as likely as not get a blank stare. Yet, in his day, he was one of the world's great railway engineers, feted by kings and prime ministers. He was eventually elected president of the prestigious Royal Society and was a founder member, and eventual President of, the Institute of Civil Engineers.

So why is Stephenson remembered and Rennie forgotten? Certainly Rennie's life is as interesting as Stephenson's. As a youth he had 'swash-buckled' with the best; wandering the back roads of Europe, battling pirates, engaging brigands and fighting off packs of wolves, all whilst hobnobbing with the great and good. Then, as an adult, he became a man who could lay reasonable claim to being one of the greatest civil engineers of his age.

According to his autobiography, he was born at 27 Stamford Street in London on the 30th August 1794, thirteen years after George Stephenson, whose path he would cross on many occasions over the years to neither of their advantages[1]. Their backgrounds could hardly have been different. Stephenson was a collier's son, whilst Rennie was the son of a famous and respected civil engineer. Stephenson was an uneducated labourer who started his working life as a child, working the 'traps', or ventilation doors, at the colliery at Wylam in Northumberland; Rennie was privately educated until he was eight years old then went on to public school, Greenlaw's academy at Isleworth, where he was taught the 'classics': geography, arithmetic, French and the rudiments of astronomy. It is a measure of the status of Greenlaw's that one of Rennie's fellow pupils was the poet Shelley who Rennie remembers as having a 'countenance rather effeminate' coupled with a 'violent and excitable temper'. According to Rennie, Shelley loved playing with gunpowder and once blew the lid off of his desk 'to the great surprise of Dr.Greenlaw (the headmaster) himself'. Rennie doesn't reveal if Shelley was reprimanded for this but we may assume that the childish prank was overlooked, being the typical high-jinks expected of a free spirited and expensive fee-paying youth. Despite his disdain for his public school his childhood compares favourably with that of the infant George Stephenson who, at the same time, was working ventilation traps in the depths of the coal mine where his father was employed.

---

1   Rennie J 'Autobiography of Sir John Rennie F.R.S. (E.& F.N.Spon, London:1875)

Rennie moved on to Dr. Burney's school at Greenwich where he made 'little further progress in anything but classics' for the two years he was there[2]. He left, aged fifteen, with the intention of joining the army and fighting Napoleon but was persuaded by his father to join the family business and become a civil engineer[13]. In the capacity of apprentice, he therefore worked on a number of his father's important construction projects, notably Waterloo Bridge, where the Thames obligingly froze over during the course of the project, and the building of Southwark Bridge where he personally supervised the cradle-to-grave import of construction material (granite) from Peterhead in Scotland. For a few years, his father's work took him all over Britain and Ireland, to oversee the survey work resulting from the proliferation of new ports and harbours which rose in tandem with the industrial revolution. In 1813 he assisted the survey of land for a potential railway or canal between the towns of Stockton and Darlington, his father's preference being for the latter. Rennie Snr. lost interest in the project when told he was expected to work alongside another famous Scottish engineer, Robert Stevenson[14]. He objected to the implication that such an eminent man as himself should ever need technical support in what was, he considered, such a trivial scheme[15].

It was of course a railway and not a canal that won the day and it was the uneducated George Stephenson who the Stockton & Darlington Railway eventually recruited as the engineer employed to build it, becoming arguably the most famous railwayman of all time. In this, like so many projects over the coming years, the Rennie family was to be thwarted by the semi-literate Geordie. In the meantime, John Rennie Jnr., after spells of work at Woolwich dock and Southwark Bridge, was 'permitted a short holiday' on the continent where he visited the scene of the Battle of Waterloo, just two months after the event took place. There he witnessed the appalling aftermath of the conflict, seeing the hundreds of freshly dug graves for the recent combatants. The post-battle tourist trade was, by then, in full swing and Rennie gathered a few mementos before hurriedly returning home in order to be present at the grand opening of Southwark Bridge, by the portly King George IV, to whom Rennie was formally introduced.

---

2   Autobiography of Sir John Rennie p.3

Rennie's father now decided that his son's education would only be complete if he personally inspected the engineering works of the great masters. He was therefore sent off on a 'Grand Tour' of Europe, in the company of his cousin General Aitcheson and Lord Hotham. After a 33 hour[16] sea crossing from Brighton to Dieppe, the party went overland via France and Switzerland to Milan, Venice and Rome. Having exhausted the possibilities for knowledge in those countries they proceeded to Greece, where Rennie acquired a Greek manservant named Demetrius. It was at this moment Rennie's swash-buckling began.

Camping one night in a former convent at Vostizza they were awoken at 2am when the door of their bedroom was forced and in stormed a band of armed pirates intent on robbery. Rennie had been forewarned of such dangers and had prudently kept a loaded pistol by his bed which he employed to good use, killing one of the attackers. After a short hand-to-hand battle the gang fled the scene. A few weeks later his party was also involved in a pitched battle with a gang of robbers working out of caves in Marathonesi. In one near-fatal encounter, Rennie was woken in his camp-fire bed by a robber who thrust a lit torch in his face and demanded money. When Rennie declined, the robber fired the gun at point blank range into his face but Rennie's luck held as the gun 'flashed in the pan' without discharging the ball, giving Rennie the chance to find his own gun and shoot the assailant.

By all accounts Rennie didn't have the best of times on his European travels. He suffered from regular bouts of malaria and a cholera infection nearly killed him. Every inn he stayed at harboured some ghastly infestation or other. The beds in particular were full of fleas and bed bugs[17]. Many of the villages and towns the party visited also seemed to be in the grip of some dreadful infectious disease, which would lay low one or all of the travelling party. His only stroke of luck was to arrive too late to board a ship from Greece to Italy, which he found to be quarantined because of a plague that wiped out all the crew and many of the passengers.

Whenever he could, he bunked up with expat Englishmen rather than being obliged to cohabit with locals who he generally had little time for. In truth he hadn't a lot of time for Johnny Foreigner at all and, in particular, the indigenous population of the Eastern Mediterranean.

He didn't like Greek people much but he preferred them to the Turks who were occupying Greece at the time, and whose soldiers he once witnessed firing indiscriminately into crowds for no apparent reason. On that occasion, Rennie waded into the soldiers wielding his walking stick to effect, whereupon the young squaddies ceased firing, apparently overawed by Rennie's *'courage and valour'*. The same walking aid was also employed to good effect when a pack of wolves attacked Rennie's party while they were camping.

In all these tales of derring-do one can't escape the feeling that at least some of this misfortune fell down on Rennie's head as a result of his own actions. At the gate to the Turkish fortress of Napoli di Romanai, for example, it was customary for all foreigners to dismount and proceed over the drawbridge on foot as a mark of respect but Rennie wouldn't countenance this and galloped over the bridge, leaving his luckless assistant Demetrius to receive a kicking at the hands of the guards. On another occasion, after much merriment and wine drinking, the Rennie party enthusiastically fired their guns in unison into the air, however the weapon of the unhappy Demetrius backfired and blew his hand off.

Rennie never seemed to get the measure of non-Anglo Saxons. Seeing a small boy trying to extricate a highly poisonous snake from a hole, for example, he gallantly leapt to the child's rescue, beating the snake to death with his faithful walking stick. The boy it transpired was unappreciative as he was an apprentice snake-charmer and the snake was intended to form part of the act.

Meanwhile, in England, Rennie's father took ill and expired soon after. His son hurried home to take charge of his father's affairs. The engineering business was bequeathed to John and his older brother George. George's expertise was in mechanical engineering whilst John's was in civils so there was no conflict of skills and a fraternal partnership prospered. One of the company's first projects was the completion of London Bridge, the project initiated by their father[18]. Because of the heavy use sustained at this river crossing, the old bridge was maintained underneath the rising structure of its replacement, a complex engineering matter that required Rennie's constant presence on site. This would have a long term cost as, during an inspection, Rennie fell several feet from a cross-beam, sustaining head injuries that affected him for the rest of his life. Nevertheless, the bridge was completed by 1830 and officially opened by King William. There was general opposition to the project, however, which was seriously over budget. Luckily, one of Rennie's strongest advocates was the prime-minister, the Duke of Wellington, whose popularity with the masses, it must be said, had suffered significantly following the Peterloo incident in Manchester[3]. However, both royal and premier patronage was enough to get Rennie a knighthood.

From 1831 onwards he would be *Sir* John Rennie.

Whilst he gained fame and honours from the completion of London Bridge, he hadn't overlooked railways. Although his father had once walked away from the construction of the Stockton & Darlington Railway, Sir John was far-sighted enough to see that railways represented the future of transport. The Rennie name was sufficient to arouse the interest of a consortium planning a railway between Liverpool and Manchester and one of their members, Lord Lowther, a Lord of the Treasury, was acquainted with Rennie from the work on London Bridge. Lowther therefore asked Rennie to engineer the new railway. Rennie agreed, provided that he and his brother would carry out the work and that, 'it did not interfere with Mr. Stephenson or any other engineer who had been previously employed'. Stephenson had only recently had a mauling at the House of Commons when he provided evidence in support of the proposed railway. The parliamentary bill he had championed was thrown out, in part because Stephenson had failed to do his homework in connection with the costs involved, but mostly because his impenetrable Northumbrian dialect and lack of schooling weighed heavily against him with the better-educated members at Westminster.

Rennie was therefore appointed 'engineer-in-chief' of the Liverpool and Manchester Railway (L&MR) and immediately commissioned Charles Vignoles to conduct a new survey. Stephenson had previously managed to upset a number of influential landowners whose land the proposed line would cross, often being barred from entry and having to carry out surveys under the cover of darkness. Vignoles's

---

3  On the 16th August 1819 a peaceful anti-poverty protest in St. Peter's Square, Manchester was attacked by government troops, killing eighteen people including a woman and child.

line avoided most of Stephenson's trouble spots but took the railway over the notorious quagmire of Chatmoss. Rennie's brother George, along with Vignoles, gave evidence in support of a new bill and the L&MR was at last given royal assent. It must have come as a blow to Rennie therefore when he was told by the railway company to collaborate with Stephenson in getting the line built. In Rennie's mind, civil engineering was a profession 'no man should be allowed to practice unless he has passed a proper examination', so it must have been a dagger in his side being expected to work alongside the ill-bred northerner whose direct involvement Rennie felt should best be 'confined to the supply of steam locomotives'.[4]

*The Victoria Dock at Hartlepool*

There was immediate friction over basics. Stephenson proposed his (now standard) gauge of 4 feet 8½ inches for the track but Rennie pushed for 5 feet 6 inches. Rennie also suggested the use of wooden rather than stone sleepers and in both matters he was over-ruled. Echoing his father before him, Rennie therefore withdrew from the enterprise, leaving the way open for Stephenson's appointment as

---

4 'Autobiography of Sir John Rennie' page 431

Chief Engineer (although Vignoles continued to carry out surveys). Stephenson's gauge, as we all know, won the day but Rennie had at least the satisfaction later of seeing the stone sleepers soon replaced with wooden ones after stability problems occurred with the track-bed over Chatmoss.

To his credit, Rennie was only too aware of the potential of railways and, over the following years, he and his brother made sure they were directly involved in one way or another in many new railway enterprises. Amongst them were the proposed London to Birmingham and Birmingham to Liverpool lines, both of which took up a lot of the brother's time and cost them a lot of money, yet neither reached fruition, at least not on the routes that the Rennies championed.

As pioneers in the new field, it was inevitable that Rennie would clash with Stephenson again and again (e.g. during the building of the Hartlepool to Haswell Railway, where Stephenson surveyed and costed the line but Rennie was responsible for the design of the accompanying new dock). It must also have been a source of irritation that the line proposed by Stephenson's son Robert, from Birmingham to London was chosen in preference to Rennie's. Their most acrimonious clash however was over the London and Brighton Railway – here, more than anywhere else, the different philosophies of the two great railwaymen contrasted most markedly.

Rennie proposed more or less a straight line from Kennington Common to Brighton, overcoming the formidable obstacles of the North and South Downs by deep cuttings, viaducts, embankments and long tunnels, particularly at Merstham and Clayton Hill. Stephenson, however, true to his principle of minimising engineering, suggested a route that wound through the river valleys of the Mole and Adur, entering Brighton from the hamlet of Shoreham to the west. This was cheaper to build but added an unwanted additional thirteen miles to the journey.

The two proposals were laid before parliament and, not surprisingly, the arguments in favour of George's railway initially won the committee's favour mainly because;

1. there were two proposed incline planes required on Rennie's line,
2. the total tunnel length on Stephenson's line was 2212 yards, whilst Rennie's was 6160 yards, with Rennie's route requiring three major tunnels at Merstham, Balcombe and Clayton,
3. the total earthworks on Rennie's line was 8 million yards and the longest embankment 1,120,000 yards, compared with 6 million and 600,000 yards respectively for the Stephenson route, and
4. the gradients were much steeper on the Rennie line.

It seems surprising therefore that it was Rennie who won the day, bolstered by a petition containing more than five thousand signatures from the residents of Brighton. Rennie's route, which took in all the major urban centres along the way, with the potential for greater commerce this presented, was a major factor, but the shorter journey time between London and Brighton must also have been a consideration for the future commuters of the South Coast.

*Clayton Tunnel near Brighton*

The line was consequently built to Rennie's specification, with the London terminus at London Bridge. The Company acquired running rights over London and Croydon Railway metals for the final section into central London. Rennie was hired by the London and Brighton Railway as Chief Engineer at an annual salary of £500, but he delegated the work to subcontractors, being very ill during most of the construction with regular bouts of malaria, a reminder of his youthful escapades in foreign climes, and occasional dizziness resulting from his fall from London Bridge. Nevertheless he found time to survey a coastal route from Brighton to Portsmouth and map out an alternative London to Brighton route via Shoreham which, ironically, followed Stephenson's proposed route along the Adur river valley for much of the way.

By 1840, the railway boom was in full swing and the Rennie brothers had direct involvement both as shareholders or engineer/surveyors in several of the new ventures[19]. George was responsible for the mechanical engineering side of the business and his company, based in Holland Street in London, supplied many of the engines used on the railways in which the brothers were involved. Amongst these locomotives were four single driving-wheel locos for the London and Southampton Railway, two 0-4-2s for the London and Croydon railway and two with the 'Brunel' 7 foot wheel gauge for the Great Western Railway. Rennie was proudest, however, of a 2-2-2 named 'Satelite', one of four supplied to the London and Brighton Railway, which reputedly achieved a commendable top speed of 60mph[20].

Rennie also crossed swords with George Stephenson over the design and construction of the Blackwall Railway between London and the East India Docks at Blackwall. It was a measure of the bad blood that existed that Rennie hints in his autobiography that Stephenson only suggested the alternative route for the line once he heard of Rennie's interest in the scheme. Stephenson's proposal, as usual, involved much less engineering but failed in the end because it meant taking a direct line through buildings, parks and cemeteries whereas Rennie's railway was raised high over the city on stone arches. Nevertheless, Stephenson's suggestion of fixed engines and ropes along the entire length of the line, to Rennie's disgust, was adopted although this old fashioned and awkward propulsion system was abandoned the moment the line joined the rest of the railway network.

By now Rennie had succeeded his good friend Sir Humphrey Davy as president of the Royal Society [21] and his services were being sought from such diverse quarters as: the admiralty, with respect to the design of warships; the construction of rolling mills for the Calcutta and Bombay mints; the draining of the Lincolnshire Fens and the construction of the breakwater at Plymouth. His close association with the nascent railways of the UK led to him being asked to lay out a railway network in Sweden, for which service King Charles XV made him a Knight Commander of the Order of Wasa (sic). He was less successful with railways he surveyed in Holland and Portugal, nor were his endeavours in Sweden without their problems as the Swedish railway committee was headed by one John Sadlier who embezzled much of the money meant for contractors. Luckily Rennie extricated himself from the enterprise before matters reached a head; leaving Sadlier to do 'away with himself at Jack Straw's tavern on Hampstead Heath'.

George Stephenson died in 1848 and his formally educated and well-spoken son Robert took over his father's affairs. Rennie seems to have had a higher regard for Robert, whose engineering achievements, particularly in respect of the engineering skills used in Robert's bridge over the Menai Straits, he evidently respected. Rennie retired in 1862 and thereby outlived George Stephenson by 26 years. Of the many railways he had some hand in, the London and Brighton remains as a living testimony to his vision. Anyone travelling today on any of the railways in Sweden or staring into the rain from the window of a 'sprinter' train travelling over Chatmoss, might reflect on the energy and ambition of the man who helped to create them. Nevertheless, the suspicion remains that, in respect of railways at least, he might have achieved so much more. If he had successes, he often, like his father before, failed to deliver the right product at the right time. His stubbornness, self-righteousness and unwillingness to work alongside lesser mortals let him down and so it is the collier's son rather than the knight of the realm that people still remember today.

**Notes to Chapter 2**

13. His father, also named John Rennie, is referred to as 'Rennie' throughout the autobiography if he is mentioned in a work related context.
14. Robert Stevenson, note the spelling, was no relation to George but was the designer of the Bell Lighthouse and grandfather of the author Robert Louis Stevenson. The names Stevenson and Stephenson often cause confusion. Robert Stephenson, for example, was both the name of George's son and that of George's brother both of whom were involved in the development of early railways.
15. Letter from John Rennie to Joseph Pease, dated 26/12/1818.
16. Those complaining about the Channel Tunnel take note.
17. He devised a trick of creating a lamp by covering a plate with cooking oil which attracted the fleas.
18. To paraphrase the nursery rhyme, the old London Bridge was falling down but couldn't be immediately demolished because of the heavy traffic it sustained. The bridge Rennie built is the one that subsequently ended up in the middle of America, reputedly because the buyer mistook it for Tower Bridge.
19. E.g. Blackwall Railway, Cannock Chase Railway, the Great Northern Railway, London, Chatham and Dover etc. Rennie was to lose a lot of money carrying out work for a myriad of small railways that subsequently went bust at the end of the boom.
20. Another member of the same class, appropriately named 'Vulture', was less praiseworthy as it spectacularly exploded on Brighton Station in 1853.
21. Bearing in mind the long running and acrimonious dispute between George Stephenson and Davy over the patent rights for their respective miner's safety lamp, this can only have fuelled the antipathy between Rennie and GS.

# Chapter 3

## *Joseph Samuda – vacuum salesman*

*Former atmospheric railway pumping station at Croydon*

In the centre of Croydon stands a forlorn boarded up building. This strange castellated structure was once used, on its current location, as a domestic water pumping station, but its origins go much further back. 170 years further back, to be exact. Back to a time when it housed air rather than water pumps, was located on a completely different site and formed a fundamental part of one of the strangest railways the world has ever seen. The railway in question was the London and Croydon Railway (L&CR) and, to find out more, we have to go back to a time when canals were king and the construction of a canal between the city of London and the town of Croydon seemed like a sound investment.

At the turn of the 19th century, roads in the country were so poor there was no effective alternative to canals for the bulk transport of goods. Canal companies, as a consequence, made vast profits. So a new waterway, linking the capital to the nearest significant population centre, Croydon, would be no exception would it?

Yes it would.

It cost a lot to build for a start, £127,000, even though it was less than ten miles long. A high proportion of this was allocated to the construction of 28 locks to accommodate the climb out of the Thames valley. The canals also needed a ready supply of top-up water and this one had none. Consequently, the company was forced to spend money on the construction of two large reservoirs and a water pumping station just to accommodate water loss. Even so, the canal still ran dry and, by the time railways came around, the Croydon Canal Company was heavily in debt, haemorrhaging money as quickly as the canal lost water. To the proprietors of the newly formed London and Croydon Railway committee, however, this same canal, or at least the route it took, had much to recommend it.

For a start, the canal took the direct line from Croydon to the city. It had easy gradients and there was little, if any, need for compulsory purchase of intervening property. In consequence, most of the major problems associated with building new railways had already been overcome. So, after a lot of haggling, the canal was purchased by the L&CR for £40,250 and construction began on converting the section between Anerley and what would one day be West Croydon Station from canal to railway[22]. The southern terminus would be built over the former canal basin at Croydon[23].

From the outset, it was never envisaged that the full extent of the canal route would be utilised by the railway[24]. For a start, the canal met the Thames at Rotherhithe, not ideal for commuting to the city. No, the idea was that the northern end would share rails already put down by the London and Greenwich Railway (L&GR), who already had a London terminus at London Bridge. This arrangement looked good on paper but proved a costly mistake. The L&GR, in their monopoly position, were able to charge what they liked for the short section of line common to both railways and space at London Bridge station was restricted, since it was shared between three railways with the inevitable disputes over scheduling. This unsatisfactory situation deteriorated to the point that the L&CR were forced to construct their own separate London terminus next to Old Kent Road in Bermondsey, near the Bricklayers Arms pub, from which their new station took its name.

Engineering of the line was entrusted to Joseph Gibbs and was not without problems. There were stability issues with some of the new cuttings, notably the deep one at Forest Hill where the embankment had to be continually remade because of subsidence. All the former canal locks also had to be removed although the timbers were recovered and used as sleepers on the new railway. Despite the problems, the line opened on time on June 5th 1839, but it had now cost four times more than the original estimate.

Much of this additional expense, it must be said, resulted from indecision by the directors in respect of fundamentals. Firstly, the Company was only partly committed to the Stephenson rail gauge of 4ft 8½in and, as a result, purchased enough additional material to allow for a transition to their preferred gauge of 5 to 6 feet. Secondly, they were undecided as to the best form of locomotion and hedged their bets, eventually using both conventional locomotives and experimenting with atmospheric railways, as we will see later. But to get the whole project quickly off the ground, and on time, they purchased eight locomotives, two from our friends the Rennie brothers and the rest from 'Roberts and Sharp' of Manchester. None of these engines were successes; indeed, the Rennie engines, 'Croydon' and 'Archimedes', performed so badly that the L&CR withheld payment for them. This had the predictable outcome that the Rennies refused to maintain them, a dispute which ran on for years, with a consequent toll on the locomotives concerned.

The railway opened on the 5th June 1939 in the guise of a train full of London dignitaries, which included the Archbishop of Canterbury and Lord Mayor. The debut train made the trip from London to Croydon in an impressive 30 minutes, with a break at New Cross on the return journey so that the passengers could enjoy a slap-up meal laid out on tables in the recently constructed engine roundhouse[25].

At the time, the railway operated over an essentially rural landscape. Peckham, for example, is described in the Company's advertising brochure as 'a pleasant village with one gas-lit street'. Between London and Croydon there were eight stations; London Bridge: New Cross: Dartmouth Arms: Sydenham: Annerly: Norwood: The Jolly Sailor and Croydon (now West Croydon). Within two years, the station at the Bricklayers Arms terminus was also being used to reduce the tolls payable to the London and Greenwich Railway (L&GR) for London Bridge. Additional stations were later added at East and South Croydon.

From the start, despite operating with only first and second class coaches, the public response was enthusiastic. In the first six months of 1840 the L&CR conveyed more than a quarter of a million passengers and the London and Brighton Railway were making overtures to the L&CR to lease their rails to act as the northern section of their own railway.

Things seemed to be on the up, yet the Company committee minutes of the time suggest that even these halcyon days were fraught with problems. The minutes are an honest, if unwisely candid, account of day-to-day issues which, for the researcher, makes for more interesting fare than the usual dry notes left by early railway committees. It seems, for example, that the staff was a surly and unruly bunch. The pages are littered with complaints from passengers about rude and/or embarrassing behaviour. One station clerk, for example, was dismissed for an act of 'gross outrage' with a female passenger. There were numerous reprimands for drunkenness including, worryingly, those attributed to train drivers, some of which resulted in actual collisions between trains. If the detail in the reports is correct, then the Company's drivers verged on the unstable[26]. On one occasion one of these worthies, a man called Jobson, on returning his engine to the shed at New Cross found his favourite locomotive parking spot already taken. In a fit of 'rail rage' he sought out the offending parkee, with whom there was an exchange of questionable language. Jobson apparently lost the argument but, being a sore loser, returned to his own engine, took a hammer from the tool box and proceeded to rearrange the features of his work colleague. In the resulting enquiry this particular incident resulted in Jobson agreeing to undergo an early form of anger management therapy whilst the battered and aggrieved other party magnanimously agreed to shake hands and make up.

The passengers on a succession of accidents resulting from employee errors were less fortunate. A series of minor rear end shunts are recorded, some of which caused actual passenger casualties. These collisions often resulted from L&CR and L&GR trains occupying the same section of line at the same time[27]. Typical of these was a collision on 30th March 1840 when an L&CR train ran into the rear of a Greenwich train in thick fog at the junction of the two operators' lines south of London Bridge. The force of the impact derailed the rear two coaches of the Greenwich train and a third Greenwich train ran into the debris. Follow-up investigations pointed accusing fingers at individual errors made by members of staff[28]. As a consequence, the railway introduced the first primitive interlocking signalling system at the London Bridge terminal, preventing two trains from occupying this common section of line at the same time.

It is an indication of the naivety of staff to the dangers of railways that one of the station employees was run over by a train whilst sweeping snow off the rails; the 'wrong kind of snow clearance' rather than 'the wrong kind of snow'.

My favourite story, though, concerns a guard who was entrusted with the care of a prize pedigree spaniel by its proud owner. Because the train wasn't busy, he shut the dog in an empty compartment while he attended to other duties. Inevitably, at the first scheduled stop the carriage door was opened by a boarding passenger, whereupon the dog made a bid for liberty. Spotting the escaping animal the guard duly abandoned the train he was responsible for and, accompanied by a railway policeman, set off in hot pursuit. Sadly the spaniel quickly outdistanced them. What passengers waiting on the abandoned train thought of the guard's behaviour is not recounted. The bereaved owner was, however, awarded £3 in compensation and the guard fined for dereliction of duty.

The London and Croydon Railway in-house police force were little better than the rest of the staff. One of their number appeared before a board of enquiry for failing to open crossing gates, resulting in a derailment that killed a guard. Another was censured for an unspecified 'act of indecency' with a female passenger, the latest, apparently, in a long line of similar complaints from passengers about the same man's lewd behaviour. He only escaped dismissal because the complainants were too embarrassed to disclose the exact details. At about the same time, a third policeman was reprimanded for using insulting behaviour to a passenger then, after the traveller complained, actually attacking him, causing bodily harm.

During the first four years of operation, scheduling of trains was constantly interrupted by engine failures. With engine breakdowns common, it is unsurprising therefore that the committee members were tempted to experiment with 'atmospheric' railways that offered the promise of trains that operated without need of locomotives.

So what were atmospheric railways?

For a short time, in the decade beginning 1835, atmospheric railways looked like the future of rail transport. The principle was simple. A vacuum was generated in a continuous pipe laid between the rails. Along the length of the pipe, a slot, kept air-tight by a sealable leather flap, connected the leading coach or wagon of a train to a piston inside the pipe. A device on the connecting rod to the leading coach lifted the sealing flap allowing air in behind the piston and the train was then propelled forward by atmospheric pressure. The leather flap resealed after the train advanced, restoring the vacuum. Older readers may recall a similar technique used to distribute cylinders of cash around large department stores.

The principal advocate of atmospheric railways was one Joseph Samuda who, along with his brother, operated a marine engineering business on the Isle of Dogs. In 1841 he published the theoretical, 'A Treatise on the Adaptation of Atmospheric Pressure to the Purpose of Locomotion on Railways', which the proprietors of the London and Croydon Railway interpreted as the way forward. Samuda must have had a convincing patter because no less a light than Isombard Kingdom Brunel, who never suffered fools gladly, began trialling the Samuda system on the Great Western Railway. It is unsurprising therefore that it was to Samuda the L&CR turned when they needed someone to build them an atmospheric railway. In support of their decision it could be argued that, if the method worked, it had much going for it, particularly in urban areas. For example, there were no polluting exhaust fumes, at least none associated with the train itself, and little noise[29]. For the operating company there was the additional advantage that no locomotives would be needed and therefore none would need to be bought or maintained. The rolling stock could also be lighter and cheaper since, effectively, they were no longer reliant on gravity to keep them fixed to the rails.

By the time the L&CR became interested the system had already been successfully trialled by the Samudas over a short stretch of track in Ireland between Kingstown and Dalkey. The directors of the L&CR went to see the experimental line in question and must have been impressed with what they saw because they immediately sought and obtained parliamentary approval in 1844 for an atmospheric railway to run on rails laid next to their existing railway from London to Croydon, with an option to extend the line to Epsom and even perhaps as far as Portsmouth, should the system prove effective[30]. The vacuum was maintained by a series of pumping stations along the way, four of which were eventually built. They were beautifully designed, looking more like fortified churches than steam pump houses. Initially it all looked very promising.

*Atmospheric railway fly-over near Norwood*

The only early problem was how their atmospheric railway would cope with junctions, crossovers and connections to conventional railways, which indeed occurred on their railway near Norwood. Their solution was to build a rail flyover, the first seen anywhere in the world. Since atmospheric trains had no locomotives that required shunting at rail termini, trains could be operated equally from either end and were therefore the fore-runners of today's diesel multiple units.

The atmospheric railway had, as noted earlier, some powerful friends, such as Isombard Brunel. However, from the outset there were obvious disadvantages, neatly summarised by Robert Stephenson when asked to brief parliament on the Dalkey railway:

1. that the atmospheric system was not an economical way of transmitting power,
2. that atmospheric railways were expensive to construct and maintain, and
3. that the system would only operate if every part of the railway was working perfectly at all times i.e. it only took a failure of one component for the whole system to fail[31].

To the above could be added the following:

4. There was no quick way of putting a train in reverse. If a train overran a platform it had to be pulled back by ropes (mostly by horses but occasionally with the less than willing assistance of third class passengers).
5. Unless the whole of the railway network could be persuaded to adopt this mode of operation there were major logistical problems at junctions with non-atmospheric railways[32].

It would also prove notoriously difficult maintaining a vacuum in such an extensive and complex system, a problem they would have found difficult to cope with even with the technology available today. With the primitive equipment of 1845 and, bearing in mind each train continually broke and remade an airtight seal as it moved along, the odds against it working efficiently were immense.

To add to this, rats quickly developed a taste for the leather sealing flaps, which had to be continually repaired or replaced[33]. The same flaps were made airtight with a sealant of tallow and wax, and this, unfortunately, melted in the summer and froze in the winter.

As a consequence of this, only one section of the atmospheric railway was eventually built[34]. It opened on the 19th January 1846, shutting down later the same day when one of the crankshafts at the Croydon pumping station snapped. Over the next month there were breakdowns pretty much every day and some of the Company's limited fleet of steam locomotives had to be redeployed on the atmospheric line just to maintain the service. The atmospheric railway section of the L&CR staggered on for a further year, now in the hands of, and at the expense of Samuda himself, but was finally abandoned in May 1847 following a fire at West Croydon that destroyed much of the specialist rolling stock. Throughout its 15 months of operation, the committee reports continued to peddle the myth that the daily problems being experienced were merely teething troubles. But it is obvious they knew that the trial had been a failure, from the fact that the Company handed the operation of their atmospheric railway back to the system's manufacturer and that they persistently refused Robert Stephenson permission to access their railway to conduct his own evaluation experiments.

*West Croydon station*

Plans to extend the system beyond Croydon were shelved pending more favourable reports of the current operation but the whole idea was finally abandoned after amalgamation of the L&CR with the London and Brighton Railway to form the London Brighton and South Coast Railway (LBSCR) in 1846. If the system couldn't be made to work on the short stretch of railway on which it currently operated then it was unlikely to be of practical use on a railway ten times that length. The complex plant and infrastructure was then decommissioned and, in most instances, so thoroughly removed that virtually no evidence of the atmospheric railway survived through to today. The one exception is the aforementioned pumping station. This was sold to the Croydon Board of Health by the LBSCR for the princely sum of £250.

The London and Croydon Railway ceased to exist as a separate entity from July 1846 but the LBSCR went from strength to strength and their rails now form an integral part of the capital's commuter chaos. The station at the Bricklayers Arms was not a success and closed in 1852. West Croydon station is no longer the main station in Croydon. That honour goes to East Croydon on the main line to the south, which was opened by the L&CR in 1841. For a time there was also a short lived Central Croydon station that stood on the site now occupied by the local authority – but that was built by the LBSCR and is another story.

**Notes to Chapter 3**

22. 'West Croydon' was originally just called Croydon. Croydon was already famous for having the first public iron railway which operated between Croydon and Wandsworth, but was horse-drawn.
23. While it was still useable, the canal was utilised to transport gravel extracted from cuttings to places along the line where it could be used as ballast.
24. What was left of the canal at Annerly was described as an 'angler's dream', as it was stocked with fish. One regular complainant was a Sydenham farmer whose pigs often strayed on to the line, due to lack of fencing, and rapidly became pork chops in the process.
25. This is believed to be the first of its kind to have been built.
26. This was not a problem solely confined to the L&CR. As early as 1828 Robert Stephenson was describing enginemen as 'unmanageable' requiring 'as much improvement as the engines'.
27. The impression given in the minutes is that the two companies didn't like each other much and hence didn't communicate effectively. One fortunate consequence was the creation of the country's first, albeit primitive, block signalling.
28. The driver of the L&CR train was reprimanded for driving too fast in foggy conditions. The 'watchman' at the junction was also reprimanded for not signalling that the Greenwich train was ahead, although how much he would have seen in thick fog remains unclear.
29. It was the latter feature that probably did for the unfortunate lady at New Cross.
30. One of atmospheric railways' greatest advocates was the engineer Sir John Rennie who had supplied the company's first locomotives. In the light of their subsequent arguments over the poor quality of Rennie's locomotives, perhaps he was an unwise council.
31. It could be reasonably argued that the Stephensons had a vested interest in conventional railways and, in particular, locomotive manufacture. It should also be pointed out that one of the biggest advocates of atmospheric railways at this time was Isombard Kingdom Brunel.
32. Compare the use of the 7ft track width by the Great Western Railway.
33. In fact, the rats were regularly sucked into the pipes and ended up forming blockages at the pumping station.
34. Between New Cross and Forest Hill.

# Chapter 4

## *Holman Stephens and the Selsey Tram*

*Holman Stephens in military regalia*

Selsey, on the south coast, is the second largest centre of population in the District of Chichester and in 1896, as it is today, was poorly served by roads.

By the end of the 19th century, the cathedral city of Chichester had lines to the capital and branch lines to Bognor Regis, Littlehampton and other coastal resorts, yet the obvious additional connection from Chichester directly south to Selsey was thought to have little commercial potential as it only served a sparsely populated agricultural area. The Light Railways Bill of 1895 provided the means for change. Under this less stringent legislation, a railway could be engineered to a lower specification and safety standards. Construction costs could therefore be minimised and a railway could effectively be built and operated on the cheap. Unfortunately 'cheapness' would be behind both the rise and the downfall of the nascent railway company which was duly created to link the world to the fishing village of Selsey. The railway's name was grander than its actuality. It was called the 'Hundred of Manhood and Selsey Tramways Company Ltd.' (HM&STR)[35].

Excited by the cheapskate opportunities created by the Light Railways Act, the Directors soon discovered that even more financial corners could be cut by maintaining a pretence that their new transport system was in fact merely a 'tramway'. By pretending to be a tramway the railway could get away with even less in the way of expensive infrastructure and safety provisions. There would, for example, be no necessity for signalling or gates where the railway crossed a highway and the rolling stock could be basic with few concessions to passenger comfort, since we are no longer talking about railways, but tramways. To further reduce costs only three of the eleven stations (Selsey Town, Hunston and Chichester) were to be manned[36]. Given these financial restrictions, the directors needed the right man to implement their minimalist vision. Their gaze quickly fell on what appeared to be the perfect candidate in the shape of the idiosyncratic Holman P. Stevens, later known, particularly to his employees, as 'the Colonel'[37].

Stephens did not come from an engineering family. His father was an artist, a member of the pre-Raphaelite brotherhood, and Stephens had only limited previous railway experience, serving his apprenticeship at the Neasden works of the Metropolitan Railway. Nevertheless, from his base at Tonbridge Wells in Kent, he forged a career which would oversee a roster of ragbag railways, using principles of thrift he intended to apply in abundance to the H.M.& S.T.R. A lifetime bachelor, he gained the title 'Colonel' only after doing service in the

Voluntary Reserve, the forerunner of the Territorial Army, where he spent most of his time recruiting squaddies. It is a pity he didn't extend his army experience to his business. On Stephen's railways no expense was too difficult to spare and no corner too awkward to cut.

There can be few railways anywhere that boasted less engineering than the Selsey Tram. It was fortunate therefore that the topography of the Manhood Peninsula, as the land between Selsey and Chichester is called, was flat and bisected by no rivers or significant watercourses, apart from a canal. As today, the peninsula is mostly occupied by farms and so, by negotiating leasing arrangements with landowners, potentially expensive compulsory land purchases could also be avoided[38]. The downside of all this was that remote halts were sometimes required, exclusively for the local landowner's use and, wherever possible, the rails followed existing field boundaries to minimise disturbance of farm layouts. Not surprisingly these arrangements ensured that the resulting railway wandered around like Branwell Brontë at closing time. The bends were so tight that speeds greater than 10mph were rarely possible. However, as it turned out, speed was not an issue. Other than the two railway terminals, population centres along the way were conspicuously avoided. Stations such as Sidlesham were actually a mile from the village from which their name derived.

Since engineering was minimised seven and a half miles of track, including eleven stations (or halts), were built for a paltry £21000, a low cost even for the time, which the company justified on the grounds that the railway was never intended for 'express speeds'. The rails were laid directly on the ground with a sprinkling of beach shingle between the rails as ballast. Stations, if you could call them that, were mostly wooden halts, occasionally topped with bus shelters.

The only significant engineered structures needed were a bridge over the tidal rife near the halt called 'Ferry Ponds', a bridge under the highway at Selsey and a peculiar, and possibly unique, drawbridge where the tramway crossed the Chichester Canal. The latter was raised by a hand operated winch, manned, at least some of the time, by the children of one of the Tramway's employees who lived in a tied cottage nearby.

*Manning Wardle 0-6-0 'Sidlesham' crosses the canal drawbridge at Hunston c.1910*

The official opening was on the 27th August 1897 when a locomotive called the 'Chichester'[39], driven initially by the Mayor of Chichester, pulled out of Chichester Terminus station an hour late, establishing a principle of timekeeping that the company would endeavour to maintain over the next 38 years.

'Selsey' was the only new locomotive purchased by the H.M.& S.T.R.[40]. The remainder were a clapped-out collection of ancient ex-works engines, mostly tiny 0-6-0 saddle-tanks capable of negotiating the Tramway's tight bends. Other rolling stock was similarly ramshackle, purchased piecemeal and painfully mismatched. In Edward Griffiths' book about the Tram[41] he reported that a holidaymaker visiting Selsey noted that the coach he was travelling in was one that had previously been written off by the London Chatham & Dover Railway, a company he knew *'never condemned anything until it condemned itself'*.

The Tram consequently gained the reputation of being the worst railway on the south coast. Trains were consistently late, if they managed to complete the journey at all, and breakdowns were daily occurrences. Children going to senior schools in Chichester always arrived late and ice cream prepared in Chichester was slush by the time it reached Selsey Beach. Consequently the railway acquired the unenviable moniker the 'Sidlesham Snail', or more picturesquely the 'Blackberry Line', as children were seen getting down from the carriages and gathering brambles during periods when repairs were being carried out. Cyclists organised races against the Tram, starting at Chichester and finishing at Selsey, and the cyclists always won.

A bad situation got worse when a major storm in 1910 caused a breach in the harbour wall at Pagham and land, which was once a natural Harbour but had been reclaimed for farmland, became a harbour once again. Since the Tram's rails were laid directly on the ground, this section of the railway was inundated by the sea and trains were unable to work over this section of the line for several months. During this period, a horse-drawn stagecoach connected the unflooded ends of track until the rails were raised on a chalk causeway above what had now become the sea bed. The expenditure associated with all this must have broken the Colonel's heart.

Surprisingly, serious accidents on the Selsey Tram were rare, the sole advantage of slow train speeds. Nevertheless, minor collisions between trains and motorists at highway crossing points were regular occurrences to the extent that eventually a by-law was enacted which required the engine driver to get down from his cab and wave a red flag before attempting to move his train across the road[42]. This was something unseen since the days of the Stockton and Darlington Railway and can have done little to enhance the Tramway's reputation.

Despite its safety deficiencies the only fatality throughout its years of operation was that of a fireman, 'Dirg' Barnes. Poor Barnes was crushed against the boiler of his engine following a train derailment north of Selsey Golf Links, with the resulting enquiry laying the blame squarely on the poor maintenance of the less-than-permanent way. According to the inspector:

*'not one sound sleeper could be found in the vicinity of the incident.'*

*Train crash north of Selsey Golf Club*

The rails were so poorly secured that pins holding rails to sleepers could be pulled out by hand. There was, understandably therefore, little pride in the Railway amongst the staff and unsurprisingly every opportunity was taken to acquire whatever illicit perks were on offer. One of the drivers, for instance, carried a twelve-bore shot gun and would stop the train in order to bag rabbits for the pot. Another had a lucrative side-line in delivering parcels to remote farmhouses along the way. He also provided an alarm call service for local residents, occasionally abandoning his train to 'knock up' late risers.

In later years the Tram's stock of locomotives were so unreliable they were relegated to goods duties and passengers were conveyed in rail buses, which were merely converted vans fitted with flanged wheels. This, on paper at least, was a clever and innovative idea for a rural railway, nevertheless travellers complained that the buses stank of petrol and were too hot in the summer and too cold in winter. Still, the rail buses did have the bonus that they usually got you there. It could be argued therefore that in the latter years of its life, the railway finally became the tram it always purported to be. It was ironic therefore that it was then that it applied for classification as a Light Railway, with the owner now Colonel Stephens himself[43].

In the heady days before the First World War, the Selsey Tram carried over 100,000 passengers each year and the future looked rosy as the popularity of Selsey as a seaside resort steadily increased. There was even talk of building an extension to Itchenor and Wittering, but traffic declined dramatically after the War and, with the death of Holman Stephens in 1932, the writing was on the wall. The Tramway now lost more money year on year and the company duly went bust in 1934.

At the end, the Tram was in a sorry state. An indication of its poor condition during those final days may be gained from a first-hand account of a journey made by railway enthusiast, Mr. W. Bishop, who, in 1933, set out with the express (pardon pun) intention of travelling from Chichester to Selsey. On arrival at the terminus at Chichester [44] he was informed that his fare wouldn't be collected until the train reached Selsey in order to give him the opportunity of deciding if he still wanted to return by train or jump ship and use the more reliable Southdown bus service to complete the return journey.

The train had barely left Chichester when the engine had to be detached from the coaches and sent on alone to Hunston to fill up with water, returning an hour later to collect the train. At road junctions, the train stopped and the driver descended and waved a red flag at truculent road users. The journey was further delayed when the train was forced to stop so that horses, which had strayed on to the track, could be rounded up. The ailing engine, leaking steam from every orifice, finally gave up the ghost at Selsey and the return was on board one of the rail buses. Not surprisingly a nominal 18 mile round trip took several hours, despite being scheduled at 35 minutes each way. Significantly, Mr Bishop was the only passenger present in both directions.

The receiver tried to persuade the Southern Railway (SR) to take on the assets, such as they were, but after examining the state of the track and the appalling condition of rolling stock the SR decided that the cost of bringing the railway up to even minimal standard was too high and, in January 1935, the Selsey Tram closed for good.

So what remains today?

Considering the Tramway vanished 75 years ago, much of its course is still traceable as most of it was laid over land which has

*Remains of canal bridge abutments at Hunston*

not since been developed. At Chichester the site of terminus station is occupied by a printing company yet the existence of the former station there is recalled by the adjacent 'Terminus Road'. Stretches of the trackway have been converted to public footpaths, particularly through the villages of Donnington and Hunston, and a token few feet of rails re-laid by The Chichester Ship Canal Trust, next to the former canal drawbridge. The station platforms at Hunston and Chalder are just about locatable, if now covered in undergrowth, and the chalk causeway embankment across Pagham Harbour is now an important public footpath through the Pagham Harbour Local Nature Reserve.

The only identifiable railway artefacts in Selsey are the cutting east of the former Selsey Bridge Station and a row of terraced houses built specifically for railway staff in Church (formerly Station) Road. At East Beach, near the site of Beach Station, there is an information board that describes the Tram. This spot also marks the start of the Selsey Tramway footpath that follows, as far as is possible, the original tramway route. Oddly enough, although 'Beach' station was the first to close in 1907, its platform was one of the last to disappear.

It wasn't until the 1970s it made way for a children's playground and reputedly still remains as the infill of the bank used for the slide.

None of the ancient locomotives, all named after local towns and villages or the subjects of classical literature, survived closure but the nameplates of 'Selsey' and 'Ringing Rock' are on display at the Colonel Stephens Museum at Tenterden and Chichester District Museum respectively. The more serviceable of the coaches from the Tram fared better, being incorporated, along with railway coaches from other railways, into holiday homes along the seafront at Selsey, where they can still, with a little effort, be seen today.

*Former Selsey Tram track-bed at Pagham Harbour*

It is possible to still get a flavour of a ride on the Selsey Tram by visiting the Kent and East Sussex Railway at Tenterden, once part of Stephens' ragamuffin railway empire. During one such visit there, with my long suffering children, the train we were on was held up for several minutes while another engine, totally unconnected to our train, was loaded with coal using a mechanical digger that completely blocked our path. In that magic moment I felt the paternal ghostly presence of Holman Stephens overseeing proceedings with an approving eye.

**Notes to Chapter 4**

35. The Hundred of Manhood is the roughly triangular shaped peninsula south of Chichester whose southernmost tip is the small town of Selsey and its western extremity, the village of West Wittering.
36. The 11 stations were in order from North to South (Halts marked (H) Chichester, Hunston, Hoe Farm (H), Chalder, Mill Pond (H), Sidlesham, Ferrry Ponds (H), Golf Links (H), Selsey Bridge, Selsey Town and Selsey beach. Ferry Pond Halt only operated from 1910 and Selsey Beach ceased to be used after 1905.
37. During World War I, Stephens became a Lieutenant Colonel in the Voluntary Reserve, the forerunner of the Territorial Army, although he never served overseas or saw active service.
38. Wayleave rights were arrangements made directly with landowners for rights of way over their land in exchange for an annual rent or other financial settlement.
39. The 'Chichester' originally belonged to the contractors who built the railway and was purchased hastily just prior to the opening since the new locomotive, the Selsey, purchased by the company was not yet ready.
40. A purpose built one off design Peckett 2-4-2 tank engine.
41. 'The Selsey Tramways' – by Edward Griffiths.
42. One lady motorist was hit by the tram in successive weeks whereupon she angrily pursued the driver as he beat a hasty retreat down the road.
43. Henceforth renamed the West Sussex Railway.
44. The Terminus Station was separate to, and immediately south of, the current Chichester Station which was originally operated by the London Brighton and South Coast Railway. The two stations were linked by a crossover line.

# Chapter 5

## *George Stephenson and the Hartlepool Railway*

*Portrait of George Stephenson*

On my bedroom wall there is an old railway poster announcing amendments to the timetable of 'The Hartlepool Railway'. Coming from Hartlepool, and previously unaware we even had a railway, I decided to find out more about it. The poster is dated 18th February 1841 and includes an engraving of a locomotive resembling an extended beer barrel on six wheels. Behind the rudimentary tender are two carriages that look like elongated stagecoaches. Even to my uneducated eye, the train illustrated belonged to the early days of railways and yet here it was promoting changes to a timetable, suggesting the railway was already well-established at the time the poster was produced. I was to discover that what I now consider *my* railway had come into existence a decade earlier than the date shown; indeed just six years after the Stockton and Darlington Railway kick-started the railway age. Further research revealed there were in total five railways which included the name Hartlepool in their title, namely the 'Stockton and Hartlepool Railway', the 'Stockton, Hartlepool and Port Clarence Railway', the West Hartlepool Harbour and Railway' and the grandly titled 'Great North of England, Clarence and Hartlepool Railway'. The 'Hartlepool Railway' referred to in my poster was the 'Hartlepool, Dock and Railway Company' (HD&R) which was the first of these.

Hartlepool, or more particularly East Hartlepool (to distinguish it from that Johnny- come-lately, West Hartlepool) had been a port since the middle ages. It has a natural harbour, protected from the ravages of the North Sea by a limestone peninsula which stands before its mouth like a clenched fist. Its advantages over nearby Teesport (at least according to the prospectus for the later Stockton and Hartlepool Railway) was:

*'…extraordinary protection and facilities, ships being able to enter or quit that place with comparative safety in all winds when not only the Tees but all the other coal ports of the east coast are closed'.*

The key word here is 'coal'. Until the 1980s, north east England was synonymous with the mineral and its many collieries fuelled the industrial revolution. It was no accident that the earliest railways in the world were built there to transport 'king' coal from the mines in Durham and Northumberland to the coast, where it could be shipped on to almost anywhere in the burgeoning empire. For hundreds of years, Hartlepool had been a busy and prosperous port but by the beginning of the nineteenth century those days were long gone. Over time, the harbour had silted up to such an extent that it was only navigable by the smallest of fishing boats. Indeed, part of the old harbour was actually in the process of being backfilled by a local farmer for his own use as a cornfield and the remainder could barely manage four foot of water, even in the highest spring tides. In the words of the engineer John Rennie, who carried out the first survey for a dock in 1832, this once flourishing town, where John Wesley had once preached to a large congregation, was '*the most primitive fishing village I ever saw*'. Nevertheless, Hartlepool, as a potential port, had obvious advantages over its near neighbours Stockton and Middlesbrough: For a start it was closer to the coalfields of north and central Durham and, being on the coast, was not subject to the extreme tidal vagaries of the River Tees.

So, at an open meeting held at the Queens Head public house in Durham on the 18th October 1831[45] a fifteen man committee was appointed with the object of:

'*...making Hartlepool a port for the shipment of coal and for making and maintaining a railway thereto*'.

Amongst those present was the then Mayor of Hartlepool, Thomas Vincent, and the meeting was chaired by Rowland Burdon, who subsequently presided over all such meetings and whose name crops up time after time in reference to novel railway schemes on Teesside.

The newly appointed committee wasted no time in applying to Rennie, now 'Sir John', to identify the '*most effectual means*' of enlarging and improving the harbour so it could be used for coal shipments. No doubt to Rennie's annoyance, at the next meeting, it was agreed that Rennie's old adversary George Stephenson would also be recruited to report on the '*line of the railway*' by the 30th of November. Annoying though it must have seemed to Rennie, it was hardly surprising that it was to Stephenson representations were made as he was now famous throughout the land, and particularly in his native north east. Unfortunately for the HD&R George's time was at a premium since, by his own account, he was already involved in '*80 such projects*', not least of which was the Liverpool and Manchester Railway, which had opened to great acclaim the year before. It is to George's credit therefore, and no small testament to his fitness, that he was able to carry out a survey of the line, with the assistance of young apprentice, Frederick Swanwick, in good time for the next committee meeting.[46]

Stephenson was summoned personally to attend and expand in detail on the cost of railway construction but failed to make an appearance, with no reason provided for his absence. Subsequent committee minutes show he was a hard man to track down. In desperation, on the instructions of the committee, Thomas Wood was tasked with finding George *'wherever he may happen to be'*.

As might have been expected, Stephenson was eventually located at Liverpool and his 'back-of-fag-packet' estimate of £133,033 for engineering the new railway subsequently accepted. A prospectus for the company was drawn up and an application made to parliament to legitimise the whole enterprise and enable compulsory land purchase powers. The Bill that created the 'Hartlepool Dock and Railway Company' was given royal assent on the 1st June 1832.

As envisaged by Stephenson, it was his intention that the colliery end of the railway would be at Hetton, in which George had a vested interest through both his engineering of the local colliery railway and the provision and maintenance of its six operational locomotives. It was proposed that as many local collieries as possible would have access to the HD&R and branch lines were therefore provided to Wingate, Thornley and the City of Durham. George wanted the line double-tracked throughout but, in the end, insufficient capital was raised and the railway was obliged to terminate at Haswell, with just one section doubled between Hesledene and Castle Eden. Meanwhile, George's reluctant colleague, Rennie, was producing ambitious plans for a double dock at Hartlepool interconnected by sea locks. Unsurprisingly his expensive proposal fared no better than George's had for double-tracking. Rennie's scheme was subsequently drastically pared down to just a single dock of 13 acres, which became the current Victoria Dock. The construction was overseen by Christopher Tennant[47]. Tennant had recently moved to the old town to live with his elderly mother and was himself getting on in years. Sadly, he would never live to see the opening of the dock whose construction he painstakingly managed.

Stephenson delegated the construction of the railway to Edward Steel. Incorporated in Stephenson's original design were two stationary engines, one to overcome the 1:34 incline at Hesledene bank and the other to control the movement of coal trucks on to wooden staiths on each side of the new dock. Other necessary engineering works included three bridges over the intervening steep-sided sandy valleys (known locally as denes), a road bridge at Throston in old Hartlepool, an 80 yard wide and thirty foot deep cutting at Hesledene and a 2 mile long, 30 foot high embankment parallel with the beach through farmland known as Hart Warren.

*Stephenson's railway embankment near site of former Hart Station*

This railway embankment would, from the day of its construction, eliminate sea views and restrict access to the beach to the residents of the expanding town, including, I should add, a younger version of myself. It remains the only significant landscape feature in the otherwise featureless housing estate where I was brought up (from a 'train spotting' perspective, it did have the advantage of raising the railway above the level of the surrounding land – effectively putting passing trains on pedestalled display).

Amongst the contractors employed on the construction of the Hesledene cutting, coincidentally, was a 'driver' also called George

Stephenson who was paid £2 13s for delivering 61 wagons of spoil. One assumes this was not in fact the famous engineer himself. Stephenson was still physically fit, but by then above the routine 'shifting of clarts' as they say in Hartlepool. It is more likely this was George's nephew, who would have been a teenager at the time and, in railway terms, just learning the trade.

The 14 mile long railway was mainly intended for coal transport so the running of passenger services was let out to other independent operations, which randomly and dangerously ran horse-drawn stagecoaches with flanged wheels over the rails. It was not long, however, before the popularity of the passenger services and associated loss of revenue forced the HD&R to reconsider the arrangement and, from 1839 onwards, they scheduled their own passenger services – as evidenced by my own wall poster.

Stations were built at such outposts of civilisation as Haswell, Heads Hope, Hesledene, Castle Eden and Hart, with Hartlepool the eastern terminus. Shortly afterwards, a connection was made to another railway, extending the line from Haswell to Sunderland while a branch was also laid down from Heads Hope to Durham city. Further shorter branches to the major colliery complexes near Thornley and Wingate would shortly follow.

The first Hartlepool railway station was little more than a raised wooden platform erected beside the track, with the ticket office consisting of the hull of a boat wrecked outside the harbour. However, a covered passenger terminal was constructed in 1844 which continued in use until 1880 and miraculously survived as a warehouse right through to the 1960s. The rest of the stations along the way, until rebuilt by the North Eastern Railway, were little more than bleak wayside halts; often a long walk from the village or hamlet that bore their name. The village of Hart, for example, is more than three miles from the station which bore its name.

The first train to travel the length of the line, on the 23rd November 1835, was a coal carrier from Haswell Colliery to the docks. It is not known what locomotives the company used in those early years, but by 1840 they owned four (The 'Lord Hood', 'Exmouth', 'Rodney' and 'St Vincent'), all of which had been purchased from Edward Bury's works at Liverpool[5]. It is in fact one of these locos which is shown on my railway poster. The fact they used Bury engines rather than Stephenson's, whose locomotives were already extensively used on the nearby Stockton and Darlington Railway S&DR and at Hetton Colliery, suggests some sort of falling-out had taken place between the engineer and his employers. Perhaps the scaling down of Stephenson's carefully prepared plans left a bitter taste, or the HD&R just couldn't spare anybody to chase around the country looking for George on the off-chance his machines didn't measure up.

*Stephenson's stationary steam engine at Throston Bridge, Hartlepool*

The early days of the HD&R were not without incident. There were several people injured in alcohol fuelled street brawls that took place between English and Irish navvies involved in dock construction, a forewarning of many such future Saturday nights in the town. There were also, inevitably, an unrecorded number of navvies, seemingly expendable, killed or injured during the railway's construction. The

---

5    The chances are that the company initially only leased steam locomotives from Stephenson.

first reported fatal 'civilian' casualty was James Pattinson of Long Bank near Birtley who was run down by a *'train of 20 wagons'* when he unwisely attempted to cross the line before checking out traffic movement. The locomotive severed his right arm and leg and he died later while being treated in Thornley.

The movement of coal trucks at the docks was controlled through an engine house designed and built by Stephenson which is now a listed building near the former site of Throston Bridge. Victoria Dock officially opened on the 11$^{th}$ December 1840 with the departure of the brig 'Britannia' from Sunderland. It was loaded with coal from Thornley and sailed out of the harbour to the sound of the bells from St. Hilda's Church, the firing of cannon. By then, however, the plans of a rival company had been advanced for a separate dock and rail-link to the western side of Hartlepool harbour, a project driven as much by the reluctance of the HD&R to share their facilities as by the vision and ambition of the founder of the future West Hartlepool, Ralph Ward Jackson .

Any rivalry between the numerous companies vying for trade on Teesside ended with their incorporation into an expanding North Eastern Railway. The Hartlepool Dock and Railway Company became part of this in 1857, four years after the N.E.R.'s inception. By then a more direct route to Sunderland and Newcastle, utilising a new line hugging the coast, was already under consideration. On the day it opened, the importance of the Hartlepool to Sunderland via Haswell line, in terms of passenger traffic anyway, rapidly declined. Additionally, as West Hartlepool was a through station on the coastal route it became the main passenger station for the two towns, the former HD&R terminus on the headland being reached via a short branch line. Direct services from the Old Town to anywhere other than West Hartlepool ceased between the two world wars. For a short time before and after World War II, a steam railcar shuttle service, known locally as the 'Tally Ho', continued to convey passengers between the two towns, but even that ended in 1952. That year the remaining stations between Hartlepool and Haswell, including Haswell itself, closed to passenger traffic, though goods trains continued using the line until the mid-sixties. The last passenger train to regularly use old Hartlepool station was a two car, diesel multiple unit that brought school children in from the collieries to the grammar school on the headland. This operated until 1964 after which time East Hartlepool station closed for good.

*Gresley Class A8 4-6-2 tank fronting a rail enthusiast's special at East Hartlepool in 1961*

Today the only section of the Hartlepool and Dock Railway still used for its original purpose is George Stephenson's two miles of embankment across Hart Warren, which forms part of the east coast line between Middlesbrough and Newcastle.

Victoria Dock is still very much in use, but the railway bridge over the road at Throston, the coal staiths and original and replacement East Hartlepool stations have long gone. The engine house, which controlled the movement of coal trucks on to the dock, still exists but is now a derelict (but listed) building near the site of the former Throston Bridge. The original line from Hart to Haswell has been converted to a public footpath, used mainly (from what I've seen) by dog walkers.

Few people I've spoken to on the headland remember the old station nor, like me, ever knew Hartlepool once had its own railway company.

The Hartlepool Dock and Railway Company is now a distant memory, a poster on my wall and an embankment where, in the fifties, wearing short pants and prescription NHS spectacles, I could be seen taking down train numbers.

*1841 Hartlepool Railway poster*

**Notes to Chapter 5**

45. Oddly enough, until recently there was also a Queens Head Public House on the headland at Old Hartlepool and this appears to have also been used for meetings of the H&DR, which is confusing when trying to work out the location of committee meetings. Even more awkward, there was also a 'King's Head' on the Headland where meetings of the rival 'Great North of England, Clarence and Hartlepool Railway' took place.
46. Frederick Swanwick drove one of Stephenson's engines, the 'Arrow', at the opening of the Liverpool and Manchester Railway.
47. We will meet up with Christopher Tennant later.

# Chapter 6

## *William Huskisson and the Rocket*

*Statue of William Huskisson MP, depicted as a Roman orator, Chichester Cathedral*

The one 'fact' that people know about William Huskisson, if indeed they know anything at all, is that he was the first person to be killed on a public railway. This 'fact' has limitations, because it is completely untrue.

The truth is that by the time Huskisson died on the 15th September 1830, even allowing for the unrecorded number of navvies who expired during the construction of the first railways, several people had already been fatally injured on the Stockton and Darlington Railway (S&DR), which had been operational since 1825. The casualties included a labourer who died in similar circumstances to Huskisson, after falling under the wheels of a wagon he unwisely clung to on the opening day. In fact, Huskisson wasn't even the first reported casualty of the Liverpool and Manchester Railway (L&MR). That dubious honour went to an onlooker who injudiciously stood next to one of the cannons that sounded off to announce the departure of the first train out of Liverpool, although it could be argued he wasn't actually present on the railway when he died. The man in question was hit by a wooden stopper (or tampion used to seal the cannon mouth when not in use) that someone had inadvertently left in the muzzle.

In fairness to the Huskisson legend that, as M.P. for Liverpool and an established and prominent figure in HM Government, he was the first true 'celebrity' to be killed on a public railway. It is also a factor that he died in an incident involving arguably the most famous steam locomotive ever built, Stephenson's 'Rocket'. The circumstances of Huskisson's demise have been extensively reported[48] but, for those who don't know, the facts are as follows:

On the opening day of the Liverpool and Manchester Railway, all the country's top dignitaries, with the exception of the King, were persuaded to take their seats on the historic first train from Liverpool to Manchester. These included Sir Robert Peel and Lord Liverpool, who were both present on a train driven by George Stephenson himself, headed by Stephenson's new engine, 'Northumbrian'. The Prime Minister, the Duke of Wellington, had been provided with his own private coach and all six steam locomotives owned by the railway company were being used because of the vast numbers of people who wanted to take part in the historic journey. The line was double-tracked with both sets of rails in use. Trains carrying less notable personages were permitted to travel in the same direction as the celebrities but on a parallel line.

Twelve miles out of Liverpool, at Parkside, Wellington's train came to a halt in a cutting. This was the first watering hole for all the engines which took the opportunity to line up and parade past the Duke's train in what was meant as a glorious cavalcade. Two such trains had already passed when the tragedy took place. The passengers in the train fronted by 'Northumbrian' had been issued with strict instructions not to leave the safety of their coaches. Nevertheless, after the second train had passed there was a long delay and a number of the VIPs climbed down on to the track to stretch their legs. Huskisson was one of these. He had not been on the best of terms with Wellington recently and, indeed, had been pleasantly surprised when the Iron Duke agreed to turn up on the opening day[49]. Consequently, he felt obliged to pick his way along the track to the Duke's coach to thank him personally for attending. After all, there was good reasons for Wellington to stay away; the Peterloo massacre, in which Wellington had been a key player, was still fresh in the memory of Manchester folk and he was guaranteed (and indeed received) a hostile reception when he arrived there[50].

Early Liverpool and Manchester Railway coaches (at the National Railway Museum at York)

Huskisson was standing on the track by the side of the Duke's carriage, chewing the fat, when 'Rocket', heading the third train, approached at speed. Here accounts vary. What *is* known is that everybody else on the track seems to have had no trouble climbing back in to their respective coaches as the engine approached, but Huskisson, appears to have panicked. He first tried to cross over the line, on which Rocket was approaching, in a vain attempt to climb the steep bank of the cutting on the far side. Finding this impossible he then tried to follow another passenger, Prince Esterhazy, who had hurriedly decamped into the Duke's coach. Even now, if Huskisson had flattened his body against the side of the coach, he might have escaped injury but instead he tried to enter the Duke's coach, holding on to the door for support. Unfortunately, it swung outwards, swinging him directly into the path of the oncoming Rocket[51]. Rocket, like all engines of the day, had no brakes. It relied on reverse gear to slow it down. This was insufficient to stop the train in the moments before impact and the locomotive struck Huskisson obliquely throwing him in front of the engine and crushing his leg. Today he might have survived. In 1830 he knew the injury was fatal and said so. His wife Emily, who was also on the train, witnessed everything.

A desperate attempt was made to save his life. George Stephenson uncoupled 'Northumbrian' and Huskisson was conveyed on a flat-bed truck to a vicarage in Eccles where he obtained medical attention but nevertheless passed away[52]. To the end, he maintained the dignity befitting a statesman and dictated (in some detail) and signed his last will and testament, right down to the donation of individual small sums to the local poor in Eartham, the village in Sussex, where he lived. The big question asked at the subsequent inquest was why he found it impossible to avoid the oncoming train when those around him seemed to have had no problem. In this, a major factor is that he wasn't well at the time. He mentioned feeling ill in a letter to his friend Lord Granville, written the week before, when he also, poignantly, talked enthusiastically about the launch of the railway to which Granville was also invited.

Huskisson, albeit something of a hypochondriac, suffered bouts of illness all his life. These included an annoying kidney complaint that meant he was often caught short at the most inconvenient times. In addition a multiple fracture also left him permanently weak in one arm. Whilst it is not clear what afflicted him on the day of the L&MR launch, given that he was sixty and that his latest unidentified illness involved periods where he suffered from uncontrollable shakes, he could well have been experiencing the onset of Parkinsonism. This might also explain his hesitation as he saw the approaching 'Rocket' and his subsequent difficulty in getting into the Duke's coach. The design of the coaches can't have helped either, being a long step up from track level. It seems ironic that his death occurred on one of the nation's first railways given that for many years he had been railways' staunchest advocate, at a time when they had few friends in the corridors of power.

Huskisson was born, in 1770, into money. His family lived at Birts Morten Court near Malvern, reputedly one of the oldest moated houses in England, and he was educated at Appleby Grammar School, a good public school, where he excelled at mathematics. When he was thirteen his family moved to France and he was still living there at the start of the revolution and indeed witnessed the storming of the Bastille. He became fluent in French, a skill that ensured that his first paid employment, would be as secretary to the British Ambassador in Paris.

Initially a supporter of the revolution, its bloody aftermath completely changed his outlook. He left France at the height of the bloodshed, when embassy staff were withdrawn, but was recruited on his return to England, as the government's liaison man, responsible for dealing with the day to day flood of refugees fleeing Madame Guillotine[53]. A succession of government appointments followed, including that of Under Secretary of War during the war with France. The first Lord of the Admiralty at this time was one Admiral Mark Milbanke, who Huskisson met at his home in Portsmouth on several occasions. It was during one of these visits that Huskisson encountered the Admiral's daughter Emily. They were married in April 1799 and moved to a large house that Huskisson owned in the hamlet of Eartham near Chichester. They lived there for the rest of their lives, though Huskisson, as an MP, was often far away on parliamentary business at his constituencies around the country, which included a spell as MP in Morpeth in Northumberland.

As an MP, he gained a reputation as a hard-working, decent man and was unusual in having friends on both sides of the House[54]. For the time, he was a forward thinking radical; an advocate of free trade who fought hard against the imposition of tariffs on imported food, especially corn, which were used to keep food prices artificially high, thus causing starvation amongst the poor. In a memorable speech to his constituents, as MP for Chichester, he once said;

*'(since) men can only live by food it does appear strange policy in the legislature of any country to pass laws for the express purpose of making provision scarce.'*

As a landowner himself, this was a brave stance to take and one guaranteed to make him unpopular with his less enlightened contemporaries at Westminster, particularly Wellington[55]. The same background also made him an unlikely candidate to champion railways. The two major opponents of railways at the time were the landed gentry, through whose land the railways needed to pass, and the canal companies, who stood to lose a monopoly on bulk freight transportation. Huskisson should have been an integral part of the opposition, as he was both a landowner and a shareholder in a number of canal projects[56]. He once even delivered a long speech in parliament in support of a canal venture in which he had no shares, the Arun and Portsmouth, which had run into financial difficulties, arguing that the government, in the interests of the nation, should provide financial help to the ailing company. He argued that canal construction would provide employment for soldiers recently returned from the war and unable to find other work. He used the same logic later to obtain funding, and hence work for the unemployed, in national and local road building schemes.

So it was no doubt surprising when, during a debate in parliament in March 1825 on the application for a public railway between Liverpool and Manchester, Huskisson was the first to step forward to argue the case on behalf of the railway company. He cited the following reason for supporting the Bill:

*'for the sake of the public good I am convinced that the railway promoters have aims above the mere accumulation of profit.'*

His reasons were perhaps not quite as altruistic as might appear at first glance, given that he was MP for Liverpool at the time and would have had half an eye on the views of his electorate. Nevertheless, his argument in favour of free trade, and against monopolies, was his defining characteristic; a characteristic for which he acquired as many enemies as friends on both sides of the House. Unfortunately, the first Bill was lost because of the unintentional demolition job done by George Stephenson in his broad Northumbrian accent, who had given the House the engineering perspective without appropriate technical back-up. As a consequence, Stephenson was openly and snobbishly ridiculed by the bill's opponents. However, it was resubmitted the following year with the technical know-how this time supplied by acolytes of Sir John Rennie. Once again, Huskisson gave a long speech in support. He also lobbied important members of both the upper and lower Houses, having recently seen cargoes of cotton rotting on the Liverpool docks due to the unavailability of transport to the Lancashire mills. This time the bill gained Royal Assent.

Huskisson followed the subsequent progress of the railway's construction with great interest. He met Stephenson at the end of August in 1829 and was taken on a tour of the more spectacular construction sites, including a walk across one of the new wonders of the world, the Sankey viaduct, followed by a locomotive hauled excursion through the newly constructed and gas-lit Liverpool tunnel. This was Huskisson's first experience of rail travel and his first encounter with steam engines; the engine in question being a Robert Stephenson locomotive called 'Twin Sisters', which was being used in the construction of the line. Huskisson told Stephenson he was 'highly delighted' with the experience, ironic in the light of subsequent events.

In October, Huskisson attended the Rainhill locomotive trials[57] and therefore it might be argued knew as much as anyone about both the delights and perils of railways. For Huskisson, Rainhill had proved an altogether more agreeable encounter with his future nemesis, the Rocket, which outperformed all the other entrants to take the £500 prize.

History seems to have labelled Huskisson only as an uncharismatic and hard-working bookworm. This may be because Huskisson was a poor speaker in parliament, often reading from notes with his head down and displaying little animation. However, none of the

*Stephenson's Rocket (at the Science Museum, South Kensington)*

contemporaneous reports or letters from people who met him socially reveal him as a 'cold fish'. Indeed, his letters home to Emily are littered with amusing character assassinations, particularly of the Duke of Wellington whom he referred to as 'Sir Gorgeous' and Lord Vansittart who he referred to as 'Old Mouldy'. In reality he was that rarest of politicians, a genuinely decent man with a far-sighted and modern outlook, who actually wanted the world to be a better place, often taking sides on behalf of the poor and uneducated against his rich and landed contemporaries. His last letter, written just the day before he died to Lord Granville, is typical in that he makes no mention of his personal health problems, but agonises over the suitability of the accommodation arrangements he had made for Granville in respect of his time in Liverpool. Given all he did for the railway cause over the previous five years, it is sad therefore that he should be remembered by the manner of his death rather than the manner of his living.

He was buried in Liverpool at the recently opened St. James' cemetery, ironically seen earlier in the week by Huskisson's wife on a guided tour of the town.

*Huskisson's home at Eartham near Chichester*

Emily hadn't moved to Liverpool during Huskisson's time in the city, preferring to stay at Eartham, where she remained a further 26 years after his death. There is a commemorative plaque to her husband inside Eartham church and a statue in Chichester Cathedral. Eartham House is now a private school and the only visible evidence of its former occupant is an 1809 cannon, taken from HMS Carron, given to the Huskissons by the Milbanke family, which stands near the entrance.

A memorial to Huskisson also stands adjacent to the track on the Liverpool to Manchester line near Parkside and there are two statues (by John Gibson), one in Pimlico Gardens, London and the other, as stated, in Chichester Cathedral. Both portray Huskisson in a toga, like some classical Greek/Roman senator; a strange representation of the man given his poor oratory skills. The Pimlico statue shows him, head bowed in thought, while that in Chichester shows him, arm extended, delivering a speech. A reflection perhaps of the different way he was viewed in Westminster and the constituencies.

As a former member of the cabinet, and colleague to the current and previous prime ministers, Huskisson was railway's first real supporter amongst the echelons of power. In the pocket of his coat on the day he died were two speeches he planned to deliver, both extolling the virtues of railways. In one of these he said:

*'The principle of railway is that of commerce itself – it multiplies the enjoyment of Mankind by increasing the facilities and diminishing the labour by which the means of these enjoyments are produced and distributed around the world'*

The sentiment apparent in these words, rather than the manner in which he died, should be the way he is remembered.

## Notes to Chapter 6

48. For example, in 'The Last Journey of William Huskisson' by Simon Garfield. The most detailed contemporaneous account appeared in the newspaper the 'Albion'.
49. Apparently, he was called the Iron Duke because he had to put iron shutters over his window following stone throwing demonstrations outside his house and not because of any reputation as a 'man of iron'.
50. The Duke of Wellington had used troops in 1822 to disperse a peaceful rally of disgruntled factory workers in St.Peter's Square in Manchester at which 11 people subsequently died. The word Peterloo was an ironic parody of 'Waterloo' for which battle Wellington was duly famous.
51. The 'Rocket', driven by Joseph Locke, was travelling at 30mph and had no brakes. It's only means of stopping was to engage reverse gear. This Locke attempted but there was insufficient time to bring the locomotive to a stop.
52. The truck was being used to convey a small band of musicians who were providing a musical accompaniment to the journey. They were forced to make the long walk back to Liverpool carrying their instruments.
53. Amongst these émigrés was Marc Brunel father of Isombard Kingdom Brunel. It is interesting to speculate how railways might have developed if Huskisson had refused his entry.
54. He was a Tory.
55. One of his opponents in a debate on the Corn Laws said 'his name was associated with odium and obloquy'.
56. Including shares in the Wey and Arun Canal Company.
57. The propietors of the Liverpool and Manchester Railway insisted on 'trialling' different locomotives before deciding on which would be used on the new railway. The trails took place at Rainhill, near Liverpool, in October 1929.

# Chapter 7

*Robert Stephenson and the standing stones of Stanhope*

*Portrait of Robert Stephenson*

Imagine you're at the top of a hill. As far as the eye can see there is nothing but empty heather-clad moor and the only sound you hear is the slough of the wind and the call of a curlew or buzzard. Far below, there is a small band of cyclists attempting the long climb up to where you stand. Heads down, the cyclists pedal, battling the gradient on a narrow strip of green that takes a direct line through the purple moor towards you. The cyclists look neither to right nor left and seemingly don't notice the half-buried marker stones, laid at intervals on both sides of the path. The stones are engraved with the letters S& DR, suggesting the presence of the Stockton & Darlington Railway (S&DR). You ask yourself what are these remnants of the world's first locomotive powered public railway doing out here in this forbidding landscape?

The answer has to do with a different railway altogether that once worked this barren landscape, but one that is equally deserving of our attention. A railway that operated over the worst terrain imaginable. A standard gauge railway that climbed to the highest point of any railway line in England and a project that very nearly bankrupted one of railway's great pioneers. The railway in question was not in fact the 'Stockton and Darlington', it was the 'Stanhope and Tyne' (S&TR).

The railway age began in the North East. It was inevitable that it did so. North east England was the perfect location for beginning the industrial revolution. It had an abundance of natural resources; most notably coal, but also iron ore and limestone. A treasure trove just awaiting the right entrepreneur, capable of providing a cost effective way of transporting the material to the customer. It would be railways that would provide the means of transport, and the docks in the north east would open the gateway to the rest of the world. As a consequence of this, within five years of the appearance of the S&DR, every natural harbour along the coast was being upgraded as a port.

Unfortunately, the wharves on the Tyne and Wear rivers were working at capacity and their proprietors were demanding premium rates for their use. Similarly, the docks on Tyneside were inconveniently located miles inland at Newcastle. It therefore seemed logical to build a new dock at the mouth of the Tyne, where access for seagoing ships was both cheaper and easier. This concept had not escaped the attention of a group of Durham business men, who included in their ranks the brothers John and William Harrison. The Harrisons were already considering a railway to connect the limestone quarries at Stanhope to Medomsley colliery. At Medomsley they originally planned to use existing wagonways to move the mineral forward to docks on the Tyne. It was William Harrison's idea to extend their railway to the small hamlet of South Shields, near Tynemouth, where a new dock would be built.

At this time, judging by the profits made by the railways, tramways and docks in the north east, the demand for coal was insatiable[58]. However, the market for limestone was less clear-cut. The most important outlet was production of lime for farmers to use as soil treatment. The brothers Harrison, however, had also noted the potential for a more lucrative use for this material in the emerging iron industry. By acquiring the ownership of limestone quarries, collieries and railways, as well as the docks through which the minerals could be exported, their company could gain a stranglehold on iron making minerals. On paper at least, the future looked rosy.

Thomas Harrison (no relation to the S&TR Directors) was given the task of surveying the line as far as Medomsley. He was soon aware there were desperate problems to overcome[59]. From the limestone quarries at Crawleyside, north of Stanhope, a tunnel under the adjacent highway was going to be needed and, beyond the tunnel, the railway would need to climb 800 feet in the space of only a mile to gain the crest of the hill. Once on the summit things were little better. From the top of the hill at Whiteleahead the railway rose and fell in a series of roller-coaster hills which culminated in the crossing of the 150 foot deep, 800 feet wide sheer-sided valley at Hownes Gill. It was obvious the engineering costs alone were going to be devastating and so it is easy to sympathise with the railway's promoters when they cut back drastically on all unnecessary expenditure.

Unfortunately, one of these essential cost-cutting measures was the purchase of land.

It had become normal practice for both railways and canals to acquire the land they needed using compulsory purchase powers once the project gained parliamentary approval. However, this was both expensive and time consuming and there was no guarantee as to the eventual outcome[60]. The Stanhope and Tyne Railway therefore took what, on paper at least, seemed a cheaper option. Individual land access agreements were agreed with each landowner along the route of the line. For the uncultivatable moorland between Stanhope and Medomsley they agreed to pay fixed annual rents of between £25 and £50/annum/mile; the same rate negotiated by colliery tramways for crossing similar terrain. This was not unreasonable and cheaper than direct land purchase. Additionally, the wayleave arrangement had the advantage that as soon as money became available, and there was soon a clamour for shares from London speculators, work could immediately start on construction of the line without protracted legal discussions[61]. It was at this point Robert Stephenson became involved, when he was appointed consultant Resident Engineer.

By 1830 the engineering skills of both George Stephenson and his son Robert were heavily in demand. George was fresh from the launch of the Liverpool and Manchester Railway and seemingly had a hand in almost all other railway projects currently on the go. Robert, a less robust version of his father, had only recently returned from an ill-fated and ill-thought-out venture to South America. With the combined experience of the Stephensons in both the Liverpool and Manchester Railway and, more close-to-hand, the Stockton and Darlington Railway it was inevitable they would be approached to oversee the construction of the S&TR. It was Robert who accepted the challenge. Since the company had no cash to spare, Stephenson agreed to accept five, one hundred pound shares in the company instead of his normal £1000 fee. This turned out to be an expensive mistake.

With the benefit of two hundred years of hindsight it is difficult to see what Robert actually saw in the project. Bearing in mind the huge financial outlay implicit in building a railway over a succession of steep hills and valleys Robert must have known it would take years before the line made a profit. His agreement to accept shares in exchange for an up-front fee therefore seems doubly strange. It is possible he may have considered this particular scheme as part of his technical education, since it was both an intellectual and technical challenge and a valuable hands-on learning experience that didn't, for once, involve his father. If so then the engineering issues presented by every inch of the railway's thirty four mile length were going to test his skills to the limit.

The first 25 miles of moor and marsh in particular were a nightmare. A dreadful combination of bog and mountain separated the limestone

*Summit of Stanhope and Tyne Railway, near Whiteleahead*

quarries at Stanhope from the colliery at Medomsley. Robert needed every available engineering skill he had acquired to deal with the obstacles along the way. It must have been particularly galling that, because of the steep gradients involved, he couldn't gainfully employ his own steam locomotives in the construction. On the moor section alone there was going to be a requirement for 10 rope hauled sections employing fixed engines and 5 self-acting inclines, apart from the aforementioned tunnel at Crawleyside. The distance between each worked incline was often little more than a few hundred yards[62]. The first two miles alone, from the quarries at Stanhope[63] to Waskerley, required the use of two enormous 50 hp stationary engines just to haul wagons up to Whiteleahead summit[64]. It was there, at a spot height of 1445 feet above sea level, that the highest point on England's rail network was about to be achieved, although the significance of this doubtlessly escaped Robert. The following half mile of open moorland was horse-drawn and a miserable, bleak, windswept and sorry place it must have been for animals to work. How any sort of service could be maintained during the deep midwinter snows prevalent in the area is a mystery.

There were further difficulties. Over the next mile, Stephenson needed another two rope inclines: one gravity driven and self-acting and the other worked by horses, now employing an invention of Robert's father, the dandy cart[65]. As a means of bulk freight transport, this method of working was painfully slow yet the biggest hurdle was still to be overcome. The line now had to traverse Hownes Gill, a wide and nearly vertical sided valley.

The obvious solution was a bridge but the engineering costs of such a structure were now way beyond the Company's means, so Stephenson came up with a cheaper alternative. At the western end of the ravine each loaded truck would be detached from the train, rotated through 90° and then loaded into a cradle and lowered down the bank on cables controlled by a stationary engine on the valley floor. The truck would then be transferred to another cradle and hauled up the other side using power provided by the same engine. Once on level ground the truck was removed from the cradle, rotated back through 90° and coupled up to its brethren ready to complete the next section, which incidentally was also rope hauled. The fact they were able to get a loaded truck from one side of Hownes Gill to the other every five minutes seems amazing under such circumstances but nevertheless would not have contributed to the speed and cost effectiveness of the railway. It is therefore not difficult to see why this ungainly, if ingenious system was eventually replaced by the long viaduct that is there today. Over the remaining fifteen miles there were four more self-acting inclines, six sections worked by stationary engines and ropes and three more by horses. No high speed records were going to be broken on this railway!

This obstacle race finally ended near where the town of Washington is today. The final nine miles to South Shields were comparatively level. And it was on this flat section that Stephenson's steam locomotives could finally be employed. There were initially seven in total, all six wheelers, and manufactured at Stephenson's locomotive works at Newcastle[66]. One even carried his name, something that might have pleased him had subsequent events not soured matters considerably.

As the railway was designed for freight traffic, it is not surprising it did not go through any significant towns or villages. Railways at that time, nevertheless, acquired their own camp following and the

Stanhope and Tyne was no exception. The line would eventually be instrumental in the creation of the towns of Washington, Consett, Chester-Le-Street and South Shields, not to mention the villages of Pelaw and Annfield Plain. The Company's decision not to operate passenger trains therefore, in the end, turned out to be one more financial miscalculation, as will shortly become evident.

The construction of the railway was completed within two years of the lifting of the first turf in 1832. It opened on the morning of Wednesday 10th September 1834 to the usual variety of razzmatazz. According to the Newcastle Journal Saturday September 13th 1834 there were:

*'rejoicings and ceremonies and the discharge of cannonades on the banks of the Tyne.'*

To the peel of church bells, the first locomotive pulled away from the staiths at the new dock at South Shields, heading west to collect a train consisting of 100 trucks of coal from Medomsley Colliery. Crowds gathered around the dockside throughout the day to welcome its triumphant return and ships moored near the mouth of the Tyne were decked out with bunting. Flags were also waved from the windows of all the houses, one of which, according to the local newspaper, spoilt the festivities by displaying 'a political slogan'[67]. By the time the train arrived back at South Shields, preceded by a brass band playing the 'Bonnie Pit Laddie', and the first coals loaded on to the collier 'Sally', a crowd of more than a thousand had assembled to witness the event. Untold quantities of brown ale was quaffed in surrounding hostelries as the party began in earnest. In the Golden Lion inn the principal employees of the railway, along with certain local dignitaries – some 121 persons in total, sat down to a feast provided at the expense of the company's shareholders. Pride of place in the centre of the proprietors' table was 'a beautiful model of a railway wagon bearing the inscription 'Medomsley Coals' which was filled to overflowing with *grapes, pines* (sic) *and peaches'*. Toasts were made to the 'British Navy' and the 'dirty faced colliers' who had made all the largesse possible. The remaining 800 plus railwaymen and other staff, who we would like to think included at least a few of the dirty faced colliers, were provided with dinner at lesser establishments after the 'Sally' had set sail[68]. A notable absentee was Robert Stephenson.

Robert had been working on various major engineering projects since completing this minor assignment (by his standards) for the Stanhope and Tyne[69]. Apart from travelling all over the country surveying and engineering new railways, he was also carrying out major improvements to the design and development of steam locomotives at his works in Newcastle. He consequently gave little thought to the few shares he possessed in the S&TR. This was a mistake.

If 1832 was a great year for the 'Stanhope and Tyne' it had been a bad year for Robert Stephenson. In October his beloved wife Fanny died after a two year battle with cancer and, unable to cope with the reminders of married life around him in his London home, Robert moved house. He had barely taken up residence in his new home when there was a fire, during the course of which Robert lost most of his most personal possessions, including the only images he had of his wife. He was therefore emotionally unprepared for the financial storm about to break around his head.

Several chickens now came home to roost. The local agreements made under the wayleave arrangements hadn't run to plan. On the open common land between Stanhope and Medomsley the landowners, principally landed clergy in the guise of the Bishop, Dean and Chapter of Durham, had charged the going rate of around £25/annum/mile as agreed. However, east of Medomsley, the next set of landlords, predominantly in fact the same people as before, spotted a good thing when they saw it, and upped the ante. For the handful of miles from Medomsley to South Shields the railway was forced to agree to a crippling wayleave rate of around £300 / mile with no possibility of renegotiating the contract for another 21 years. The enormous engineering costs were not yet recovered and the daily handicap race over Waskerley Moor meant portage was uncompetitive compared with similar suppliers of minerals. With the company losing money the difficult section of line between Stanhope and Medomsley was first to close. Even so, the company was still expected to pay rent to the landowners because of the machinations of the wayleave arrangements. The canny Harrison brothers saw the writing on the wall for S&TR and jumped ship, selling their shares at a healthy profit but leaving the remaining shareholders to field a subsequent flood of bills.

Acquiring debts faster than it gained customers, the company nevertheless continued to borrow heavily, ploughing some of the capital raised into another expensive railway project, the 'Durham Junction Railway' (DJR), a project designed to connect the Hartlepool Dock and Railway Company to the S&TR. Despite leaving most bills unpaid, the company nevertheless contrived to pay 5% dividends to its shareholders using money borrowed to fund the DJR, a method later adopted with enthusiasm by George Hudson. Sadly the outcome was the same as that which caused the demise of the 'Railway King'.

Things couldn't continue in this way. Within months of the celebrations at South Shields, the company had accumulated debts of more than £400,000. Belatedly the S&TR applied for rights to run passenger services but the wayleave landlords objected[70]. They received a further blow when the Brandling Junction Railway (BJR) opened for business. By linking Gateshead and Monkwearmouth via South Shields, freight business was stolen from the S&TR. Given what was going to happen, it was ironic therefore that Robert Stephenson was one of BJR's directors.

Year-on-year the situation worsened until in December 1839 Stephenson was presented with an unwelcome Christmas present. He was handed a staggering bill for debts incurred by the S&TR[71]. To say he was stunned was an understatement. He had long since written off the shares in the company he had been given and thought that was the end of it. He was wrong. The S&TR Company was never incorporated and therefore had no limited liability. In consequence, each of its owners was deemed individually responsible for any debts incurred. Since Robert was almost certainly the best known, and hence the most accessible shareholder from whom to seek redress, he was the first port of call for the long line of creditors. Robert rushed off to consult a lawyer and was made aware of the deep financial pit into which he had fallen. It turned out that he stood to lose most of his personal fortune, and all for the sake of a £500 stake in a company he had no continued interest in. No wonder he told his friend Edward Cooke:

*"Ordinary rascality bears no relation to that which has been brought into play in this affair."*

Robert's lawyer, however, saw a way out. An extraordinary meeting of the shareholders was held in which it was agreed the company would cease trading and a phoenix company produced from the ashes. Fresh capital could then be raised on behalf of the new company. All money previously invested in the Stanhope and Tyne would, nevertheless, be lost. To finance his share in the deal Robert had to sell his stake in his beloved loco works at Newcastle, albeit to his father. By such means the former shareholders of the S&TR somehow managed to clear the Company's debts. This was too late, however, for the S&TR which was wound up on the 5th February 1841, having operated as a working railway for less than seven years. What assets remained were transferred to the phoenix company, now called the Pontop and South Shields Company (P&SS). Not surprisingly the only section of the line operated by the P&SS was the profitable eastern section.

The remaining western section, between Stanhope and Consett, was sold to the Derwent Iron Company and surprisingly reopened in1842 for the sole purpose of transporting limestone from Stanhope to a new ironworks at Consett. To facilitate transportation of iron onwards from Consett, another line was constructed from Waskerley to Crook, which connected directly to the Stockton and Darlington Railway. It was therefore only a matter of time before the S&DR absorbed this section into their expanding empire. Under the new arrangements, the Waskerley to Stanhope section of the former S&TR line continued to work until 1965. Operating under an Act of Parliament, compulsory purchase of land was now possible so the Railway was no longer under the thumb of greedy landlords. Despite these powers, land access was severely restricted, giving the S&DR just a few feet of land on either side of the track, the edges marked by boundary markers.

Here, at last is, the answer to the question of those mysterious marker stones on the moor. The unique inclines Stephenson built at Hownes Gill continued to be a major restriction. Even when the S&DR dispensed with Stephenson's ingenious method of moving trucks across the valley, and reverted to a more conventional standing incline, the best speed achievable for shifting trucks over the valley was still painfully slow. Consequently, the S&DR eventually bit the bullet and constructed the handsome viaduct that dominates the Gill today.

*Hownes Gill viaduct, near Consett*

The eastern section, no longer encumbered by the time taken for wagons to cross the moor, now also started to make a profit. It went on to provide the main export route for iron and steel from Consett to Tyne Dock for another hundred years, ending with the closure of the steel works in 1980.

Little now remains. The western section is a cycleway and footpath. The limestone quarries at Stanhope are still used, although the stone the quarries produce is now taken away by road. The remains of rails and sleepers from the former S&TR can still be seen on the margins of the stone quarries, some being used for fence posts. Stanhope itself is the terminus of a heritage railway that operates over a few miles of the former S&DR through the Upper Weardale valley. One forlorn, graffiti-covered 'slag' truck stands near where the rail junction into the works were once located. Much of the eastern end of the railway is built over, but the branch line to Durham, extravagantly built using borrowed money, now forms part of the east coast main line between Newcastle and York. Rowley, one of the stations subsequently built on the S&TR by the S&DR, has been removed and rebuilt in the nearby industrial museum at Beamish.

And finally, raising their heads over the top of the heather, are the S&DR standing stones. A reminder to walkers and red-faced cyclists of a strange switchback railway that once bested the surrounding wild moor but very nearly bankrupted George Stephenson's only son.

*Stockton and Darlington Railway stone boundary markers on the moor above Stanhope*

## Notes on Chapter 7

58. For example, within the space of one year following the opening of the Stockton and Darlington Railway, the traffic of coal on that railway increased from 4,000 tonnes/month to 11,000 tonnes/month.
59. Harrison was to go on to become the Chief Engineer and General Manager of the North Eastern Railway.
60. George Stephenson had recently fallen foul of the system himself when his application on behalf of the Liverpool and Manchester Railway had been rejected in the House of Commons and he was to have many similar battles with the elected members over the other railways he engineered or proposed up until the day he died.
61. Eventually there were 50 'proprietors' of this railway, mostly London based speculators.
62. Tom Rolt, in his biography of the Stephensons ('George and Robert Stephenson the Railway Revolution'), mentions a locomotive-hauled section immediately south of Hownes Gill but there appears no reference to this in other studies of the line, most notably in Tomlinson's 'North Eastern Railway'.
63. The railway terminated high above the village of Stanhope. It is not the same line as the heritage railway that still serves the town, which was part of the Stockton and Darlington Railway's Weardale branch.
64. One of these stationary engines, still in working order, can be seen at the National Railway Museum at York where demonstrations of its operation are made on the hour each day.
65. Dandy carts were reputedly the idea of George Stephenson. On the downhill sections of the line the horse was given a lift, and a presumably welcome rest, in a specially adapted cart fastened to the tail end of the train.
66. All were 0-4-2s named after directors and other people associated with the railway (including the 'Thomas Harrison' and the 'Robert Stephenson') apart from the 'James Watt'. One of these is depicted on the covers of the company's ledgers and accounts.
67. It read 'Earl Grey and No Revolution' which hardly qualifies as insurrection.
68. Minus the two men and a boy killed on the 1st of May when the first section of line from Stanhope to Annfield Plain opened to traffic and a wagonload of workers broke away from its rope fixing on the Weatherhill incline and careered out of control to the bottom of the hill.
69. These projects included the eventual major trunk routes of London to Birmingham and London to Brighton.

70. Passengers, for a nominal payment, had been unofficially hauled over the route in open wagons since the line opened and occasionally an 'unofficial' coach would be attached to trains although one assumes that passengers made their own way across Hownes Gill.
71. Why Stephenson, as a minor shareholder in the company, should have been singled out isn't clear, although Rolt speculates that his fame and reputed wealth went before him.

# Chapter 8

*Thomas Richardson – steam to nuclear power*

The parish of Castle Eden consists now, as it did at the beginning of the 19th century, of a handful of houses, a church and a pub, at the junction of highways linking Sunderland to Stockton, and Hartlepool to Durham. A 'blink and you miss it' type of place, you could say. For a few years, however, during the reign of Queen Victoria, it was the centre of industry. In those glory days the village boasted a colliery, an iron foundry, a cotton mill and a (sadly missed) brewery[72]. The transformation of this rural idyll to industrial heartland owed much to one man, Thomas Richardson.

Born in the village on the 9th July 1796 to Thomas and Mabel Richardson, his father was a labourer on the Earl of Durham's estates and Thomas's early working years were spent as a forester and, later, timber manager for the wealthy landowner. At the heart of the village was a small blacksmith's forge which belonged to a family friend, George Ord, and it was there that young Thomas acquired the skills of iron working. It was a good time to be a metal worker: the first railways were under construction in the north east, on the back of an insatiable demand for coal, and railway pioneers, George Stephenson and Timothy Hackworth, lived, or at least worked, nearby. Indeed Hackworth's Stockton and Darlington Railway locomotive works was a mere twelve miles away, at Shildon.

At the age of 25 Richardson married Isabella Heslop[73] and their son, Thomas Richardson (II)[74] was born the following year. In 1832, with the help of two brothers, he leased land at Castle Eden for the site of a metal foundry. The site was not chosen by accident. At that moment an army of navvies were in the process of constructing a railway from Hartlepool to the collieries and the location chosen for Richardson's foundry was where the Hartlepool Dock and

Railway (HD&R) line crossed the Sunderland road[75]. The Chairman of the HD&R at the time was Rowland Burdon[76]. Burdon was also the main landowner at Castle Eden and Thomas Richardson's former employer. Since the railway and docks needed copious supplies of timber and metal, it was inevitable therefore that it was to Richardson's nascent iron foundry and timber sources that Burdon turned.

*The site of Richardson's foundry at Castle Eden*

Consequently, Richardson's foundry flourished, to the extent that within five years he was considering an even bolder venture, shipbuilding. Castle Eden foundry had supplied iron fittings for boats, notably anchors and Richardson consequently had all the contacts he needed when boat construction began. However, he now needed a site for a boatyard. The recently reinvigorated port at Hartlepool, five miles away, seemed the logical choice. With his close friend, Joseph Parkin, he leased land in the town behind the ancient sea wall and close to where a ferry across the harbour mouth to Middleton operated. It was here that Thomas built his first wooden hulled ship, for the Hartlepool General Shipping Company, which he unsurprisingly named 'Castle Eden', the first ship officially recorded as being built in the town.

The site chosen for the shipyard was not ideal. To launch the boat they needed to dismantle part of the town wall and then rebuild it afterwards so it is unsurprising that, within three years, the company was sold and Richardson began searching for a better prospect[77]. By then he had become involved in the opening of the Castle Eden and Wingate Collieries. What this involvement amounted to is unclear, but his works at Castle Eden would have been ideally suited to supply any construction materials the new enterprises needed. What we know for certain is that by 1851 the York, Newcastle and Berwick Railway (YN&BR) was paying Richardson for coal supplied to them from Castle Eden Colliery. So, even though he was not the colliery owner, he seems to have handled the purse strings. It was the iron foundry at Castle Eden, however, that would provide his contribution to the history of railways. Thomas began making steam locomotives and stationary engines there.

It is unknown who the first customer for one of his steam locomotives was since only foundry 'job' numbers were ever maintained[78]. These start at Job No.101, with numbers 101 to 103 perhaps related to the manufacture of stationary engines. Job numbers 104 and 105, however, were definitely for two steam locomotives supplied to the HD&R. These 0-6-0s, with their four foot driving wheels, were acquired by the Hartlepool Railway in 1840[79]. Proximity to this railway meant the foundry was ideally placed to carry out H&DR's manufacturing and maintenance work. Richardson's first involvement may indeed have been to act as maintenance contractor for the Bury engines the HD&R already used. By 1843 all HD&R locomotives were being both maintained and manned by staff belonging to Thomas Richardson and Sons[80]. His Company also acted as subcontractors, operating passenger services and mineral traffic on the Hartlepool Railway's behalf.

Since Richardson was effectively running the HD&R in 1840, the locomotive illustrated on a Hartlepool Railway poster for 1841 may also be one of those made the previous year at Richardson's foundry[81]. His design would have been 'borrowed' since there is no evidence that Richardson ever used any of his own. Richardson seems to have specialised in 0-6-0s but the illustration in question shows an 0-4-2 and is therefore likely to be foundry works number 107.

Five locos were produced the following year in Richardson's foundry for the HD&R, and all were 0-6-0s. Richardson's works continued to manufacture engines with the same basic 0-6-0 pattern over the whole of the decade to local railways, including the Stockton and Hartlepool Railway, the Newcastle and Darlington Junction Railway and, in particular, the York, Newcastle and Berwick Railway, who eventually acquired all existing HD&R stock, much of which had been built or rebuilt at Richardson's foundries[82]. Of the drawings showing Richardson's engines that survive, all the locos bear more than a passing resemblance to Timothy Hackworth's Shildon output of the time. Indeed, the engine Richardson used for many years to move ships' ballast around Hartlepool docks seems identical in appearance to one built by the firm of Hackworth and Downing for the Stockton and Darlington Railway, again consistent with the idea that Richardson only 'borrowed' other people's designs for his locomotives.

*The site of Richardson's foundry at Castle Eden*

By the end of the decade, Richardson was running two foundries. In 1847 he had leased the premises of the former Hartlepool Iron Works at Middleton, on the opposite side of the harbour from his former ship-building site, and from there he began making marine engines.

Many years earlier he had built the first marine steam engine in his original works in Hartlepool for a wooden hulled sloop called the 'Isobel' (named after Richardson's wife), and used 'Isobel' as a means of experimenting with novel engine designs. Marine engine production would soon become the principal focus of interest for the Richardson family, although they still retained an interest in actual shipbuilding through a company founded by his son Thomas and brother John in 1844. In 1851 Richardson's produced the first 'commercial' marine engine, for a paddle-wheeled tug boat which was built in Sunderland.

For a few brief years, the locomotive works and shipbuilding and marine engineering works existed side by side, with the Castle Eden works dedicated to the locomotive side of the business. During this time, several locos were produced for the collieries[83] as well as works engines for the Consent Iron works and at least 8 engines manufactured on behalf of the Robert Stephenson Company[84]. In 1855, the Hartlepool based foundry and shipbuilding works launched the first iron clad steam driven ship ever built in the town, a collier called the 'Sir Colin Campbell'. By then, Thomas Richardson, the former woodsman from Castle Eden, had died. He passed away in 1850 and his son, also Thomas, took over the various businesses. For the next seven years Richardson factories continued to manufacture, rebuild or repair railway engines, however, the Company's principal interests now lay elsewhere.

The lease on the Castle Eden foundry expired in 1853 and was not renewed. A handful of colliery locos were built, or rebuilt, over the following four years at the Hartlepool foundry and some of these survived in various forms until the turn of the century. In 1900 Richardson's amalgamated with Christopher Furness and Westgarth and Co. to form Richardson Westgarth, in which guise they continued until the mid-nineties. The original firm of Thomas Richardson & Sons had by now manufactured nearly 1000 marine engines. The engineering knowledge the Company gained was used to good effect into the nuclear age when steam turbines for both Trawsfynydd and Dungeness B nuclear power stations were provided by 'Richardsons, Westgarth and Co.'.

THE OLD BALLAST ENGINE

*Richardson's dock ballast engine*

The family connection with Hartlepool and Castle Eden is now broken. The engineering works and foundry at Hartlepool closed in 1994. Most of the extensive works there were demolished, although a few buildings still stand at Middleton, some of which are currently used by waste disposal contractors. Of the foundry at Castle Eden, the only remaining evidence is the name on a row of terrace houses which occupy the original site. The second Thomas Richardson went on to become MP for Hartlepool and his son, the third Thomas Richardson, was made Alderman of the town and given a knighthood in 1898. Considering his importance to the growth of the town, it is a pity that there are no portraits of Thomas Richardson senior in either the new Maritime Museum or the Hartlepool Art Gallery, although the local museum does have portraits of his many less illustrious descendants. At its height, Richardsons Westgarth employed 7500 people, most of them recruited locally. The closure of the company consequently accelerated the industrial decline of the town, following hard on the heels of the demise of the shipbuilding industry and the steel works, a series of blows from which the town is now only starting to recover.

Nevertheless, those surviving members of the Richardson family can reflect on a fine legacy which included a bit-part in the creation of the world's first railways. If the Richardson family had a relatively minor role in steam locomotive history this was, nevertheless, significant at a time when the future of railways was still in doubt. Richardson's contribution to marine engineering over more than 150 years is, however, second to none.

*What remains of the Richardson Westgarth works in Hartlepool*

### Notes to Chapter 8

72. The Nimmos Brewery no longer exists, alas, alas. Any draught beers bearing the name 'Castle Eden' are now brewed at the Cameron's Brewery in Hartlepool.
73. Isabella is probably the daughter of Robert and Mary Heslop, listed in the births in the parish register for 1801. Their daughter was, however, christened 'Mary'. This would have made her 19 at the time of her wedding to Thomas, which would seem to fit. Robert Heslop was employed by Rowland Burdon, the local landowner and future chairman of the Hartlepool and Dock Railway Company.
74. The name itself causes confusion, particularly in respect of early railways. Within the Richardson family itself there were three generations of Thomas's all of whom were involved in the engineering works. In addition a Thomas Richardson (no relation) was one of the founding members of the Stockton and Darlington Railway.
75. Later the A19.
76. The Burdons, like the Richardsons, weren't very original in the choice of names for their first born. This particular Rowland was the fourth generation to use this Christian name and there were to be two more before the blood line faded out. It is unclear where Thomas Richardson, who must have had little collateral of his own, obtained the money to lease the land at castle Eden and build a foundry. It is possible to speculate that Rowland Burdon could well have been his benefactor – if not, he may have acted as guarantor to any loan obtained from the bank.
77. The section of wall dismantled was where the ferry used to run between old Hartlepool and Middleton and can easily be identified since the stones used now appear darker than the adjacent yellow limestone – see photograph.
78. A complete list of the engines believed to have built, rebuilt or repaired appears on pages 547 to 548 of 'British Steam Locomotive Builders', painstakingly compiled by James W. Lowe.
79. Lowe lists engines referenced by ex-Castle Eden Foundry works numbers 104 and 106.
80. The name of the company is a misnomer in that the first Thomas Richardson only had one son, although he did have two daughters.
81. See drawings of Hartlepool Railway engines in 1841 and Stephenson engines from mid 1830s.
82. This is where the 'build' listings start to get confusing because at least some of the engines accredited to Richardson's as new builds could be rebuilds of existing engines and therefore double counted. A further

locomotive, built for the Stockton and Darlington Railway in 1845, is mentioned by Peter Hogg in 'Richies' although this doesn't appear in either Lowe or the British Locomotive Catalogue 1825 to 1926.
83. For example, Lambton Colliery, Marley Hill Colliery, South Hetton Colliery, and Wingate Colliery.
84. These were built in 1846/7 when the capacity of Stephenson's own works was stretched beyond production limits.

# Chapter 9

## *George Graham's Journal – A catalogue of errors*

*Photograph of John Graham taken towards the end of his life*

Driving a coal train on the 1st July 1828, John Cree, an employee of the Stockton and Darlington Railway (S&DR), pulled up next to a halt at a watering stop at Aycliffe Lane Station (now Heighington), County Durham. He was at the controls of arguably the most famous locomotive ever built – Locomotion No.1. Whether the boiler was nearly dry at the time we will never know, but the moment cold water entered it, it detonated, killing Cree and seriously injuring Edward Turnbull, his fireman, who was manning the water pump. Turnbull would display the scars of the explosion, a mottled black shrapnel tattoo, for the rest of his life.

Cree's was the second such fatal accident in weeks on a railway that had only been operating for three years. Steam hauled public railways were in their infancy and ground-rules for their safe operation were yet to be defined. In consequence, safe working was largely trial and error, the problems exacerbated by the terms of the railway's authorisation, which allowed anyone ready to pay the appropriate tariff the freedom to run their own trains over railway metals, regardless of age, experience or competency. Unfortunately, at least at the beginning, this was invariably the option chosen by S&DR customers, since it was cheaper and viewed as a logical extension to the in-house transport arrangements already employed. This combination of unrefined safety procedures, unforeseen hazards and irascible and uncontrollable railwaymen, led to a number of bizarre and hair-raising incidents.

As mentioned previously, William Huskisson, who died on the opening day of the Liverpool and Manchester Railway, is often incorrectly cited as the first person killed on a public railway. In fact, there were already many recorded fatal railway accidents before Huskisson fell beneath the wheels of Stephenson's Rocket. The S&DR, being the oldest of the public railways using steam locomotives, had the longest history of such incidents, all of which were attributed in their informal records as being down to negligence, human error or incompetence.

It was not until the appointment of S&DR Traffic Manager, John Graham, in 1831 that accidents and dangerous incidents were even afforded any attention, mainly from the perspective of damage to rail infrastructure and timetables that such incidents incurred. Some of these records have survived and provide a fascinating insight into the haphazard safety arrangements employed on the world's first public railway. They came to light through the agency of John's son, George, who was born the year John Cree died. Like his father, George had a long and productive career, with both the S&DR and later the North Eastern Railway (NER), and when he retired he was approached by his former employer and asked if he had anything of note to say about those early days.

Luckily for us he did.

In 1896 Harold Oxtoby, on behalf of the North Eastern Railway (NER), sat down with George to compile a journal of George's memories[85]. These hand-written notes are now kept in the National Archive at Kew, and offer a glimpse of the astonishing way the railway operated in those early years. Amongst the fascinating, if disturbing, incidents reported are those relating to 'dandy carts'[86].

Until the 1850s, mineral traffic on the railway was transported by a mixture of steam and horse-drawn trains, with the latter in the hands of private operators using their own horses and wagons. Used to driving carts on turnpike roads, the drivers were a law unto themselves and, unfortunately, were mostly left to their own devices until such time as a dangerous incident threatened the railway company's finances. Despite the S&DR's claim to fame as the first public steam railway, most trains in the early years consisted of three or four coal wagons and a horse. On downhill sections, which effectively meant most of the line between colliery and coast, no additional motive power was needed other than gravity so, when the horse's haulage services were no longer required, it trotted along beside the train while the wagons freewheeled, the driver astride a primitive braking wagon. This was a waste of the horse's energy, and the company's engineer, George Stephenson, suggested using low sided open-backed carts, located at the end of the trains, where the horse could rest when otherwise not required. The process involved unhitching the horse from the head of the train at the top of the incline then leaving it to wait for the dandy cart while the wagons trundled by. A bale of hay was kept in the dandy cart as an incentive for the horse to leap on board but, in fact, a nice rest was the only encouragement the horses really needed[87]. In principle, the concept was sound and problems only arose because of the manner in which the carts were handled.

Just months following his appointment as Traffic Manager in 1831, John Graham reported that: On the 4th August 1831, Thomas Sanderson, an independent driver, partial to the odd drink, crawled into the dandy cart at the rear of the train to sleep off a lunchtime skin-full, leaving his horse to toil uphill unsupervised with a half-dozen loaded coal wagons. The incline took its toll and the train slowed to a crawl. Following on was a sixty truck locomotive hauled train, headed by the Timothy Hackworth designed engine, 'Globe'. This struck the back of the dandy cart, derailing itself and blocking the line for two hours (also no doubt seriously disturbing Sanderson's beauty sleep). His nap cost him a 5/- (25p) fine from the railway company.

*A Stephenson 'dandy cart' at the National Railway Museum, York*

The following November there were two similar incidents on the same day (the 24th) involving dandy carts. In the first, driver Thomas Leng, failed to slow his wagons sufficiently on a descending gradient at Myers Flat. His intention was to link his wagons to the train in front, which was slowly being braked down the incline, but he lost control of his train. The loaded trucks careered down the hill, crashing into the dandy cart belonging to the train in front. The collision ejected the resident horse which fell down the adjacent embankment. Leng was fined 2/6 (12½p) by the S&DR. It is not noted what happened to the horse. Later the same day, driver George Snaith, in an almost identical incident, also failed to control his wagons on the long descent into Darlington. The runaway wagons once again collided with the dandy cart of the train in front, this time not only dumping the resident horse onto the trackside but also Snaith himself. The horse, in this instance, emerged unhurt but retribution was enacted on Snaith who not only broke his leg but was fined 5/- .

As with Thomas Sanderson, alcohol was often a contributory, if not the major, factor in many of these incidents. Early the following year, driver William Ogle and assistant George Hodgson staggered bleary-eyed from an alehouse in Shildon to take charge of a horse-drawn train. Shortly after setting off, indeed on the first downhill section, the dandy cart parted company with the rails. Neither Ogle at the head of the train nor his drunken cohort Hodgson, acting as brakeman, noticed the cart bouncing wildly along behind, tearing up the track bed. Oblivious to the damage caused and inflamed by drink, Ogle ignored warnings from the public and refused to give way to anything in his path. An oncoming horse-drawn train was forced to reverse and take shelter in the first available siding despite having right-of-way, and when Ogle encountered a long coal train pulled by a Robert Stephenson built engine, 'William IV', he refused to pull over to let it pass, in total disregard of S&DR standing orders. A brawl ensued, which involved the drivers of both trains, to which Ogle's drinking partner Hodgson enthusiastically contributed, the fight pursued even on to the footplate of the William IV. When overpowered by the sober S&DR men, Ogle turned out to be a sore loser and attempted to derail the 'William IV' by placing a rail and chair on the track in front of it. All these indiscretions were too much even for the normally laissez faire S&DR and Ogle and Hodgson were referred to a magistrate for prosecution.

Beer-fuelled belligerence was a common occurrence. Two months earlier, Robert Sanderson, William Vasey and Michael Howe abandoned the train they were in charge of for thirst-quenching activities at a pub in Spring Gardens. Since the railway was single line worked in this section, this effectively curtailed all movement on the line and both men were fined 2/6 (12½p). Driving a train seems to have been thirsty business since Sanderson and his mates were not the only railwaymen who took on alcoholic refreshment at inappropriate moments. Rather than complete the delivery of coal to a waiting collier-boat at Newport, near Middlesbrough, eight railwaymen abandoned their train of wagons and disappeared into a pub for several hours, preventing the boat they were meant to be loading from sailing. They were each fined 10/- (50p). Drunken behaviour wasn't confined to independent operators. John Lydle, a company man, was reported in July 1833 for abandoning his locomotive at Heighington station, having sank one too many at the local alehouse. His train had to be taken on to Stockton by his two firemen. Rather than receiving praise for the otherwise commendable reluctance to operate a locomotive whilst sozzled, Lydle was dismissed.

Single line working was a major factor which led to conflict. Non-company drivers saw no reason to give ground to anyone else, particularly paid employees of the S&DR. This was unfortunate since the railway in the beginning was mostly single line, with just a handful of passing places. In one incident, drivers William Myers and John Pears, in charge of a horse and four wagons, refused to pull over to allow a steam train to pass. Oblivious to the abuse hurled upon them, they prevented the train following on behind from passing for 4 miles and were both fined 5/- (25p). On stretches of single line, before the advent of signals, the S&DR devised a procedure whereby marker poles were set out at regular intervals at the mid-point between passing places. Thus lengths of track up were divided into sections of variable length, the distance between the poles and the passing places only limited by visibility. Right of way was given to the first train to pass each pole and thereby enter a nominated protected section. Once a train passed the marker trains coming in the opposite direction were, therefore, required to reverse to the nearest passing loop and permit the train with right of way to pass. This was red-rag-to-bull provocation to the independents working on the railway. Ralph Hall, with a horse drawn train, despite seeing an oncoming loco pass the relevant marker post, before Hall's train entered the same section, steadfastly refused to give way, resulting in a stand-off between the drivers and a pitched battle. In Hall's case, this was only the latest in a series of similar offences and his permission to use the railway was withdrawn.

In a way that seems astounding today, members of the public regularly rode with the driver in the engine's cab, or even, for a small back-pocket contribution, perched astride loaded coal wagons. This led to many serious accidents, in which the S&DR refused to acknowledge responsibility. Both independent and S&DR railwaymen treated trains as their personal taxis. In fact, the practice was 'officially' against S&DR rules and action was occasionally taken against their own employees for infringements. Typical of these, was an incident in November 1831 when a company driver, Ed Corner, was fined 2/6 (12½p) for allowing a man and woman, who Corner had only just met, accompany him on the footplate of his engine. It seems that action was only taken in this case because it followed hard on the heels of a warning he had only recently been given regarding a similar breach of company rules.

Having criticised the drivers, it is only fair to say that the railway itself was hardly a paragon when it came to safe working practice. Two of their apprentices had to have limbs amputated following shunting accidents during the year 1835. George Graham personally witnessed some appalling failures in equipment and work procedures which caused accidents or 'near misses'. For example he records that wagon wheels could only be adequately oiled while the train was moving. This was achieved by choosing a moment when a train was rolling along slowly on level ground. The driver and fireman would then step down from the footplate and run along the length of the train on either side, applying oil to the wheels. An unwanted side-effect of this was that the resulting reduction in friction increased the train's speed, which often obliged the footplate-men to either scuttle back to the engine and leap aboard as the train accelerated or, if the wagons were loaded with coal, jump on to the end wagon and work their way forward along the tops of the trucks 'Keystone Cops' style.

No locomotives, or tenders, were fitted with brakes. Stopping an engine meant engaging reverse gear. On a falling gradient this was hard to achieve as it involved rotating the gear wheel in the opposite direction to that of the wheels on the engine, at a specific rate of four cab wheel turns for every loco wheel revolution. This was almost impossible after dark since the driver and fireman could not see the loco's wheels and no lamps (or other distracting signals) were allowed on the footplate. In consequence, on a section of line between Eaglescliffe and Darlington, the S&DR engine 'Rocket' was involved in a collision with a horse and four wagons driven by one John Usher. Luckily, the horse was the only fatal casualty of the incident.

It appeared from later testimony that both trains had been invisible to the other right up to the point of impact[88]. As a result of this, when driving at night, firemen took the precaution of carrying lengths of tarry rope which they lit from the grate and waved from the footplate to advise their presence to oncoming trains.

Engine breakdown was a regular occurrence on the S&DR into the 1850s and beyond. One wet November afternoon George Graham (without an accompanying fireman – a practice also against company rules but, nevertheless, commonplace) was despatched from Shildon with a relief engine for a passenger train after the train's loco broke down between Middlesbrough and Redcar. Moving parts on the engine had seized due to the presence of sand blown from the banks of the nearby River Tees. The defective engine was barely moveable but Graham gamely set off back with it to Shildon despite having to stop every couple of miles and crawl underneath the engine to apply oil to keep the loco moving. The engine had also been using salt water in the boiler, a practice which caused violent surging, with water regularly expelled from the chimney. The fluctuations in boiler pressure meant the engine's water gauge was unreadable. Graham therefore also had to stop at regular intervals to try and estimate how much boiler water he had left. During one of many such water stops, near Darlington, Graham was approached by a colleague who asked for a lift back to Shildon. Over the next mile, with stops and starts for oil and water checks and water spewing out of the chimney, Graham's passenger took fright and jumped ship, despite there being many miles to go to their destination. He told Graham it was safer walking home in the dark, in a strong wind and heavy rain, than run the risk of being blown to pieces.

It was not uncommon for railwaymen to work long hours. In October 1844 George Graham and George Scott (an 'Outside Locomotive Foreman') was sent by William Bouch, Locomotive Superintendent of the S&DR, to collect four engines they had purchased from the Midland Railway (MR) at Derby. The two men were put up overnight in a hotel and, being unused to the bed provided at the inn, Graham found himself unable to sleep. It transpired that the engines were not ready for collection the following day and the next evening there was a shindig at the pub which kept Graham awake all through the night. It wasn't until 2.30am on the third night, by which time Graham had been awake for nearly 70 hours, that the men were informed that the locos were available for collection. The men eventually set off back to Shildon at 4am with one of the newly purchased engines hauling the rest. The two S&DR men had been assigned a 'conductor' by the MR who was instructed to travel with them until daylight and set them on the right way, since neither of the S&DR men knew the line. The conductor duly stayed with them as far as Clay Cross, then told the S&DR men to find their own way from then on. In Graham's words, 'We did not know one yard of the line, or any signals between Clay Cross and Darlington.' Graham was driving and Scott was firing and the leading engine was leaking badly and losing steam. At Altofts Junction, where the lines from Derby to Leeds and York went their separate ways, they chose the wrong arm and proceeded several miles along the line towards Leeds before discovering their mistake. In panic, they reverse shunted the dead engines back to the junction, praying that they didn't encounter a train coming the other way. By now Graham could hardly keep his eyes open and in fact could later recall little about the subsequent journey, which eventually took 18 hours. Graham at the end had not slept for nearly 3½ days.

It must be said, Graham himself could also be dangerously maverick when the need arose. On Good Friday, 1861, Graham was asked, by John Dixon the S&DR Consulting Engineer, to conduct a trial to calculate the rail friction of newer and heavier coal wagons, as opposed to the usual two ton chaldrons, on the steeply descending gradient between Barnards Castle and Darlington. The idea was to allow loaded wagons to freewheel unhindered down the 1/82 gradient and see if the trucks gathered sufficient momentum, over the 10 miles of descent, to propel the train up the rising gradient into Darlington without further assistance. As a consequence of this, 8 loaded trucks were attached to a guard's van housing, with twelve brave men required to assess speed and stability. Their secondary purpose on the train was to act as emergency brakemen, should that prove necessary. Since it was a Bank Holiday, the line between Barnard Castle and Darlington was devoid of rail traffic and all level crossings between the two towns were closed to road vehicles for the day. Graham stationed himself at the 10 mile point, near Winston Station, to watch proceedings. It was not expected that the experimental train would reach 60mph, since this was the fastest achievable even by express trains at the time. However, by the time the train hurtled through Winston station, it was already travelling at 73mph and was still accelerating, with 2 miles of descent remaining. From his vantage point on the embankment near the station, Graham watched the train lurch alarmingly round a bend, with each loaded wagon waddling, in Graham's words, 'like a duck'. The train barely slowed as it climbed out of the river valley.

Meanwhile, in the guards' van, attempts to apply the brakes had failed and the men trapped within were committing their souls to God, convinced they were going to die since the speed attained by the train exceeded the limit of the gauges installed to measure it. The situation was worsened by the fact that the only window of the guard's van was so encrusted with coal dust blown off the wagons it left the men pondering their fate in complete darkness. In the end, the men's luck held. The track levelled out on approach to Darlington and the train slowed sufficiently to allow the brakes to work. The wagons were diverted into a siding outside Darlington station where the runaway finally came to a halt. A red-faced Graham reported the outcome to his manager the following day, who was, by all accounts, 'very severe on the matter' although what alternative result might have been expected from this dodgy experiment is by no means clear. Illegal high speeds weren't restricted to experiments. Despite the low

*Stockton and Darlington Railway footplatemen on the 0-6-0 'Middlesbrough'*

speed limits imposed by the company for freight traffic, this didn't stop drivers using any possible means to reduce journey-time – which, coincidentally, increased liquid refreshment breaks. In 1833 a driver was reported for running a steam hauled coal train non-stop from Shildon to Middlesbrough, a feat believed at the time to be impossible, since the engine was obliged to take on water at regular intervals. The solution to the mystery was that the driver had devised an ingenious way of working. On falling gradients, when approaching the main water stopping points at Darlington and Eaglescliffe, the driver uncoupled the engine and ran the locomotive forward light to the water stop, leaving the rest of the train to coast down the hill under gravity, under the control of the fireman. Assuming the loaded wagons could be braked sufficiently, he would then accelerate his engine away from the water tank until he reached the same speed as the moving wagons, whereupon he would acrobatically couple them up to the engine and proceed as before. On Saturday nights, the same driver regularly dropped his fireman off at Darlington, where he lived, so his mate got an early finish. He then drove and fired the engine himself the remaining eight miles to Shildon. Since the Hackworth designed locomotives being used at that time had a 'U' tube type boiler, which required a driver at one end and a fireman at the other, it is unclear how he did this, but no doubt it involved some heroic acrobatics.

In truth, positive innovation was neither likely nor expected from a workforce composed, at first, of unskilled men drawn from a largely agricultural community. At the end of every month between 1830 and 1840, the S&DR was besieged by vagrants, described by John Graham as 'of weak intellect', unable to get work anywhere else and seeking employment on the railway. They gathered around the Shildon site in informal camps, sleeping under canvas or utilising S&DR trackside cabins. On one occasion, rather than offend them unnecessarily, John Graham told the group who presented themselves for work that he could not take them on because they didn't own their own watches which, he advised them, were an essential tool for railway workers. The following month the same people turned up again, this time carrying watches they had mysteriously acquired whilst simultaneously, one assumes, causing appointments missed in the surrounding countryside. The poor quality of support staff was often used as an excuse for bad work practice. Robert Pickering, a driver summoned before the S&DR

board following a series of misdemeanours, blamed all his problems on a succession of 'unsatisfactory' firemen he was given to assist him. When asked to name any of them he could only recall their nicknames 'Badger', 'Bullet', 'Baggy' and 'Buck'. Unimpressed by the defence offered, his superiors gave Pickering the 'boot'. However his bosses were equally guilty of unsafe practice. Railways were still considered money-spinning playthings by autocratic owners and shareholders, like a giant train set which others might use at a price but for whom first dibs always went to the men at the top. In 1838, for example, the S&DR applied to the Clarence Railway (CR) for permission to run passenger trains over their lines. A meeting was arranged with the CR's General Manager, Mr. Child, which was attended by George Stephenson and John Graham representing the S&DR. Child invited Stephenson and Graham to join him for breakfast. He lived just outside Stockton and arranged for the train the two men were travelling on to stop outside his home which was located right beside the railway track. While fellow passengers on Stephenson's train fumed, Child and his guests sat and dined on a hearty breakfast, resuming their journey only when they finished their meal.

Fortunately for us, most railway companies learned from their mistakes. However the S&DR, being first in the field, were reluctant to discard practices established on inception and which they considered had served them sufficiently well. The use of horses, for example, continued on the S&DR long after it was discontinued by other railways. Although the company advised their customers not to use their own drivers and wagons – even offering financial incentives to do so – horses continued to plod along their lines right up to the time when the company was taken over by the North Eastern Railway. This may, perhaps, have had something to do with S&DR's insistence that their rival CR use only horses and not locomotives on the part of the S&DR line the CR leased. Dandy carts also survived until 1856. Decent signalling did not appear on their railway until after the NER takeover, although they experimented with various prototypes, including the dubious practice of displaying a lighted candle in a Station Master's window to indicate to passing train drivers when the road ahead was clear.

Given the above tales, you might think it surprising that the outcomes of these incidents weren't more serious. However, the incidents related from Graham's journal, are not intended to provide a definitive, or even partial account of accidents and incidents on the S&DR. This would be a much bigger, more harrowing and often heartbreakingly tragic affair. Rather, I hope, this provides a glimpse of daily life on railways at a time when there were few controls. Nevertheless, it might reasonably be argued that the pattern for positive change was established by railway people such as John Graham, who at least was recording what was going wrong from day to day. Things could only get better.

*Stephenson's 0-4-0 'Locomotion' at Darlington's 'Head of Steam' museum*

**Notes to Chapter 9**

85. This was certainly a common problem at the time. A dry boiler was identified as the cause of a similar explosion a few weeks earlier.
86. According to Oxtoby's journal, George Graham is reputedly the last person to drive 'Locomotion No. 1' as a working engine (in its rebuilt form following the boiler explosion).
87. In Victorian times, the word 'dandy' apparently was applied to anything new or innovative and used liberally in non-railway contexts, as in the expression 'fine and dandy'. The nearest modern equivalent is 'cool'.
88. Most depictions of dandy carts, e.g. the illustration in Tomlinson's book about the NER and the model at the NRM at York, show the horse standing. In fact, this proved impossible for the horse if the train was moving. Horses therefore sat back on their haunches, as dogs do. Horses became so familiar with the principle that, as Tomlinson reports, if the cart wasn't attached to the train an untethered horse would attempt to climb into the end wagon. It is interesting that the death of horses belonging to their customers was usually more costly to the railway company than the death of their staff. Until the creation of a mutual insurance scheme in the 1840's railway families had little recourse if the main breadwinner was killed or seriously injured in the course of his employment.
89. This was not Stephenson's famous Rocket, from the Rainhill trials at Liverpool, but another engine with the same name. Compensation paid by the S&DR for injuries to horses belonging to their customers invariably exceeded that paid to the families of their own injured railwaymen. This led to the creation of Mutual Societies which provided financial support in such cases.

# Chapter 10

## Samuel Sidney's rural rides

'David Copperfield' had just been published, the Crystal Palace was under construction in Hyde Park ready for the Great Exhibition and, of lesser fame in 1850, Samuel Sidney was setting out on his fact finding railway trip around England[89]. The resulting book, 'Rides on Railways', appeared the following year, making 2011 the 150th anniversary of its publication.

A more appropriate name for the book might be 'Rides on Railway' since Sidney restricted himself to the lines owned and/or operated by the London and North Western Railway[90]. It remains unclear why Sidney wrote it. There are no stated aims and the text wanders around nearly as much as the author did during the book's research. If it was meant as a promotional guide book, it would more likely have deterred people from visiting the places he describes than otherwise. At best it was a Victorian equivalent of today's 'Rough Guide', meant for a growing band of rail travellers; a pocket book identifying the 'must see' and 'must avoid' locations the unwary traveller could encounter along the way. Nonetheless, the book is fascinating as it tells you as much about Sidney himself as the places he visited.

So who was he?

He was born Samuel Solomon in 1813 to a Birmingham doctor. The family was sufficiently well off to pay for him to attend law school and Sidney began his working life as a solicitor. Whilst practicing law he started writing magazine articles about agricultural and rural issues, with special emphasis on his lifelong love – hunting. Nobody, it seemed, enjoyed killing animals as much as Sidney. The rapid growth of railways gave him the freedom to travel the length and breadth of the country to indulge his questionable pursuit. This heart-warming interest in the slaughter of our native fauna was beneficial to him later in life when he became the hunting correspondent for the Illustrated London News. Despite having no background in either agriculture or railways he considered himself an expert on both and, by the time he set off on his epic journey in 1850, had already written extensively on these unrelated subjects[91].

Three years prior to 'Rides', he had written and published 'Rough Notes on a Ride over the Track of the Manchester, Sheffield, Lincolnshire and other railways' which was, in many ways, a model for the later book. As with 'Rides' it was designed as a pocket book and, like 'Rides', it contained various 'amusing' anecdotes about hunting, farming and industrial practice. There were two principal points of difference. Firstly, 'Rough Notes' was couched in the passé language of the 18th century and secondly, and more importantly, it was mind-numbingly boring. Although only three years separated the publication of each book, the differences were startling. The earlier book painfully adopted the affected structure of eighteenth century prose with 'f's replacing 's's in the text. It begins modestly, with the line;

'I am not an agriculturalift' and then sets out Sidney's railway credentials thus;

*'Railways and their capabilities for creating developing and conveying traffic have long been my ftudy'.*

'Rough Notes' was more about farming practice than any 'ftudy' of railways and contained very little to stimulate the Victorian rail buff. Indeed it is unclear as to who, other than a dedicated few in the agricultural community, would have struggled through its 200 or so dull pages. 'Rides on Railways' was a different matter. It was aimed squarely at the emerging breed of railway travellers and sold in great numbers in the Victorian equivalent of W.H Smiths.

How did Sidney view the England of 1850? To find this out we must travel north with him on an expedition from London to Leeds and hear his description of the failings and inadequacies of the towns and townies he encountered along the way.

*The entrance to Euston station as it would have appeared in Sidney's day*

He started his journey at Euston Station, just a short cab ride from his home in Fulham. The station then stood in splendid isolation, a long walk from the nearest road (and hansom cab or bus access). It was a testament to the rapid development of railways that only forty years earlier the same site was being used by Trevithick to demonstrate the novel principles of steam locomotion to a largely disinterested metropolitan audience. The engraving of Euston, in Sidney's book, shows the station entrance as a miniature Marble Arch, engraved with the already defunct name 'London and Birmingham Railway'. Sidney notes that the imposing façade was not matched by the station itself which was,

*'a plain unpretending stucco structure with a convenient wooden shed in front barely to save passengers getting wet in rainy weather.'*

The first class waiting room he reported was also:

*'dull to a fearful degree and the second class room is a dark cavern with nothing better than borrowed light.'*

'Comfort', according to Sidney, 'has been sacrificed to magnificence'. Those unfortunate travellers sat on their luggage whilst checking the departure boards at Euston today might wonder as to what Sidney meant by the term 'waiting room'. Apparently this was an innovation of the Victorian age and consisted of a heated room where passengers sat in moderate comfort to await the departure of their train. In fairness to Network Rail, some things have improved since 1850. The station is now totally enclosed so passengers need no longer get 'wet in rainy weather'. Nevertheless, there are precious few places to sit these days and the building itself, unlike the Greek temple of Sidney's day, is stunningly unimaginative. Comfort has now been sacrificed to functionality.

Sidney goes on to describe the flotsam and jetsam waiting to board the parliamentary train[92]. London parliamentary trains were, apparently, the worst in the country,

'having a large proportion of swells out of luck' mingling indiscriminately with the hoi polloi. Sidney prudently allowed the cheap rate Parliamentary to depart and then boarded a first class coach on a 'mixed' train heading north. Mixed trains consisted of both goods and passengers. The 'goods' included horse and carriages chained to open flat-bed trucks; a questionable practice even at the time. Sidney describes an incident whereby a certain Lady Zetland's horse drawn carriage, containing her Ladyship and attendants, caught fire from sparks from the engine. Her ladyship and her entourage, it appears, were obliged to beat a hasty and no doubt undignified exit.

Nonetheless, it seems that mixed trains had their plus points since, to the author's delight, they allowed packs of horses and hounds (and even such sporting quarry as deer) to be transported to distant hunts. Sadly, Sidney complained, it was:

*'a class of traffic to which few of the railway companies have paid much attention.'*

One is left wondering why.

The first stop was Camden. In Sidney's day, Camden was a mega-centre of railway activity. In fact, its growth had been both recent and rapid. Dickens visited Camden shortly before Sidney when he reported: [93]

'The first shock of an earthquake had, just at that period, rent the whole neighbourhood to its centre. Traces of its course were visible on every side. Houses were knocked down; streets broken through and stopped; deep pits and trenches dug in the ground; enormous heaps of earth and clay thrown up; buildings that were undermined and shaking, propped by great beams of wood. Here a chaos of carts, overthrown and jumbled together, lay topsy-turvy at the bottom of a steep unnatural hill; there, confused treasures of iron soaked and rusted in something that had accidentally become a pond.

Everywhere there were bridges that led nowhere; thoroughfares that were wholly impassable; Babel towers of chimneys, wanting half their height; temporary wooden houses and enclosures, in the most unlikely situations; carcases of ragged tenements, and fragments of unfinished walls and arches, and piles of scaffolding, and wildernesses of bricks, and giant forms of cranes, and tripods straddling above nothing. There were a hundred thousand shapes and substances of incompleteness, wildly mingled out of their places, upside down, burrowing in the earth, aspiring in the air, mouldering in the water, and unintelligible as any dream. Hot springs and fiery eruptions, the usual attendants upon earthquakes, lent their contributions of confusion to the scene. Boiling water hissed and heaved within dilapidated walls; whence, also, the glare and roar of flames came issuing forth; and mounds of ashes blocked up rights way, and wholly changed the law and custom of the neighbourhood

In short the yet unfinished and unopened Railroad was in progress; and from the very core of all this dire disorder, trailed smoothly away, upon its mighty course of civilisation and improvement.'

Luckily, by the time Sidney arrived the railway was firmly established, although still the main focus of Camden Town. As now, and in Sidney's day, virtually all the raw materials needed to keep the capital working needed to be imported while, at the same time, other goods such as fish from Billingsgate and vegetables from Covent Garden were heading in the opposite direction, for destinations throughout the UK. Much of the traffic to and from the north was centred on the extensive Camden goods depot [94]. Since refrigeration or other forms of long term food storage (apart from the salting of meat) was unknown, live cattle and sheep were transported to the capital and then slaughtered close to the markets where the meat would be sold. The animals were delivered to Camden in the middle of the night and herded into large, crowded, evil-smelling pens next to the marshalling yard. From there the distribution of the animals, plus any other imported goods, was divided between horse drawn vehicles, handling the lightweight goods, and canal long boats that dealt with the bulkier materials, such as coal. The barges used the nearby Grand Union Canal (or Regents Canal as it was known to Sidney).

*Camden roundhouse today*

Sidney spent some time watching the shunting procedure in the marshalling yard. Scrawny horses hauled laden trucks out of sidings and moved them into unloading bays. There was a slight downward incline from siding to bay which caused the trucks to gain momentum. This meant the horse pulling it had to be detached at the last moment whereupon it leaped out of the way to avoid getting crushed as the truck rolled to a stop on the buffers. The horses, which carried out the shunting operations, as well as dealing with local deliveries, were housed in subterranean stables (or 'vaults', as Sidney describes them) next to the railway, and frightful, badly lit, squalid places they apparently were. Locomotives, being more expensive to purchase and maintain, fared better than horses. They were 'stabled' in two recently built sheds – a 'long house' for passenger/express engines and the now famous Camden Roundhouse that was used for mixed traffic and goods engines[95]. Nearby there was a workshop for loco repairs and a carriage maintenance depot.

Sidney would be pleasantly surprised by Camden today. There is even still much he would recognise. The stables (although the horses are now only represented by bronze statues) have metamorphosed into a lively underground (in all senses of the word) street market which rivals the best Istanbul or Tunis has to offer. The Roundhouse, as most will know was, until recently, a music venue, echoing during the Swinging Sixties to the sounds of Pink Floyd and Jimi Hendrix. It is currently undergoing renovation as a trendy Arts centre. If Camden still retains many of the features Sidney noted (the Roundhouse, the stables, the working canal), his next stop, Harrow on the Hill, would be unrecognisable to him.

The station he alighted at (now 'Harrow and Wealdstone') lay, a mile from the nearest habitation, in the midst of 'rich pasture' where cattle were fattened up prior to transport into the city. There is little evidence today of the 'rich pasture' amongst the sprawl of suburban housing and shopping malls. Harrow School, which Sidney describes with the airy vagueness of one who never went there, is still there but is an incongruous vestige of upper class privilege in an otherwise urban desert. Sidney travelled north through an essentially rural landscape. Pinner, future birthplace of Reg Dwight (or Elton John as he later styled himself), gets barely a mention, nor does Watford or Kings Langley, both of which were then quiet hamlets that were a long walk from the stations that bore their name. It is therefore to Berkhampstead that he next turns his attention.

From his description, he would find the area near the station little changed from the day he stepped on to the station platform all those years ago. The view of the grand Norman castle on the east side and the Grand Union Canal on the west side of the railway would be immediately recognisable to him. Indeed, the most significant difference would be the station itself. The original station is now a builder's yard and its replacement is a hundred yards further to the north. The illustration of Berkhampstead Station in Sidney's book shows a Victorian family happily strolling along the track, presumably having just left the stationary train shown. They are blissfully unaware of any danger. Fast-forward to today and, even if they had the sprinting prowess of Usain Bolt, they would be lucky to escape a collective mangling under the wheels of the 100mph plus expresses that incessantly shake the awnings of the station canopy. To the passing 21st century rail traveller, Berkhampstead is now just a short-lived roar and blur.

*Berkhampstead station from an original engraving (as reproduced in Sidney's book)*

Aylesbury was Sidney's next port of call. He summarised it neatly:

*'With a few exceptions the farming is as bad as it can be, the farmers miserably poor and the labourers ignorant to a degree which is a disgrace to the resident clergy and gentry.'*

Its saving grace, if indeed it had one, was that the town,
*'was the centre of Baron Rothschild's (stag) Hunt.'*
Little improvement today then.

What a contrast with the good burghers of Woburn who, in Sidney's words included those:

*'finest of all modern characters – the English country gentlemen, educated, yet hearty, a scholar and a sportsman, a good farmer and an intelligent considerate landlord....'*

Sadly, the other stations on the branch line he followed to Bedford, such as Ridgemount and Ampthill, had:

*'nothing about them to induce a curious traveller to pause.'*

Bedford itself was only notable for the charity hospitals and schools:

*'where a small tradesman can send his child to a free school, where even books are provided.'*

What education was provided for children whose parents were too poor to be 'small tradesmen' is not reported. From Bletchley, Sidney took a branch line train to Oxford, noting along the way that Banbury 'is more celebrated that worth seeing' and Bicester is the 'centre of capital hunting country'. He added that the women of Bicester 'make a little bone lace' without indicating whether or not he thought this a good or bad thing. Thus, having dealt with Buckinghamshire, he arrived at Oxford.

In 1850 the Great Western Railway was still trying to gain access to the city[96]. At this time, the only rail link to the City was the LNWR Buckinghamshire Railway branch line that Sidney used. This was something of which Sidney approved as he applauded the resulting protection of the architecture of the ancient city from the ravages of pollution which would be consequent upon the arrival of those upstart seven foot gauge giants from the West Country. The Oxford station which welcomed Sidney was Rewley Road. This station was closed by British Railways in1951 but was not dismantled until 1999 when it made way for a business school for the University. This redevelopment led to a local outcry, including an occupation of the site by protesters, the outcome of which was the transfer of much of the building to the Buckinghamshire Railway Centre at Quainton Road.

Sidney devotes more pages to Oxford than anywhere else on the route with an emphasis on the apparent wonders of the University. Even Oxford undergraduates receive praise as being:

*'not more random and extravagant than any set of young men of the same age would be if thrown together for two or three years.'*

Sadly, in Sidney's time, 'the fashion of drinking (had) gone out to a great extent'. There was little wrong it seems with Oxford in Sidney's day. Even the Hunts were highly commendable, the Oxford 'Meet' being:

*'well worth exhibiting to a foreigner – where in scarlet and in black and in velvet caps, in top boots and black-jacks, on twenty pound hacks and two hundred guinea hunters, (the) finest specimens of young England are to be seen.'*

One assumes Hunts in those days were not followed by today's entourage of jobbing builders in Range Rovers looking to make the right sort of contacts.

From Oxford, Sidney retraced his steps to Bletchley, heading for Wolverton where more obviously his affection for railways and railway people are apparent. Wolverton was a town created specifically to serve the railway. In 1850 everybody who lived in the town worked for the railway company other than the tradesmen who fed and clothed them. Locomotive construction at Wolverton was new when Sidney dropped by; just a handful of locos had been built prior to his visit[97]. Sidney waxes lyrical on what he discovered in the foundry and machine shop:

*'If our painters of mythological Vulcans and sprawling Satyrs want to display their powers over flesh and muscle they may find something real and not vulgar among our iron factories.'*

As he watches sections of locos being fitted together there is an impression of awe, and even humility, not evident anywhere else in the book. The engineers he noted were educated craftsmen of 'above average intelligence' who 'all are gentlemen'. It is amongst the molten metal, clanging hammers and screeching lathes, the saws and the guillotines, that Sidney at last finds something both modern and classless to admire:

*'It is not mere strength, dexterity, and obedience, upon which the locomotive builder calculates for the success of his design, but also upon the separate and combined intelligence of his army of mechanics.'*

Amen to that.

Wolverton today is a poor cousin of its Victorian antecedent. The station alone, in 1851, housed 775 staff, of whom 4 were overlookers (managers?), 9 were foreman, 4 draughtsmen, 15 clerks, 32 engine drivers, 21 firemen and 119 labourers the rest being mechanics and apprentices. Sidney reported that:

*'a few passenger carriages are occasionally also built as experiments under the direction of the engineer J.E.McConnell'.*

Wolverton under McConnell would become the largest carriage works in Great Britain and the first to use electricity for lighting and drive machinery. The royal train was built there in 1903 and one of its descendants was fitted out there as recently as 1977. More than 200 engines were built at Wolverton between 1845 and 1863.

The grand works Sidney reported on have now, however, shrunk to become a 'Railcare' service depot and much of the land on which the works stood has been sold off. Where Sidney once marvelled at the splendour of locomotive building, there are now shoppers wheeling supermarket trolleys.

Nowhere else in 'Rides on Railways' does Sidney become as emotionally involved as he is at Wolverton, surrounded by the heat of foundry furnaces and the clash and clamour of industrial hammers. His journey, however, was just beginning. From Wolverton he headed north over the dramatic viaduct that crosses the Ouse, which is illustrated in his book. The Ouse Valley landscape is little changed, judging by the drawing. The area was recently quarried for gravel but the quarries are worked out and landscaped and now a haven for wildlife (I counted 12 herons in the space of 200 metres).

*Wolverton viaduct from an original engraving*
*(as reproduced in Sidney's book)*

Unfortunately, I made the mistake of parking a mile away from the viaduct, near the busy A5, in the sort of off road car park where cars pull up alongside, with their engines running, and the drivers smile at you winsomely and unnervingly. Still, the walk to the railway was very pleasant, through quiet water meadows where locals from nearby Milton Keynes walk their dogs. The footpath headed east under an iron aqueduct supporting the Grand Union Canal. The first and best view I had of Sidney's viaduct was from just beyond the tunnel under the canal. Most of the arches are now buried in undergrowth or obscured by trees but, other than the presence of a caravan park, the valley, now part of the Ouse Valley Park, would still be recognisable to Sidney.

According to Sidney, the embankment that leads to the viaduct was built from alum shale. This must have contained a high proportion of combustible material because it caught fire soon after it was laid down, luckily before the first train travelled along it.

Trains today scorch along the viaduct at three figure speeds, travelling between London and Birmingham; Sidney headed onwards and upwards, at a slightly slower speed, to Peterborough via a now defunct line that ran through Oundle. The first stop was Northampton whose 'silly' populace Sidney berates for resisting the overtures of the railway companies and hence missing out on the opportunity of being a 'main station' on the London to Brighton line, rather than the shoe manufacturing backwater it had become. Despite this, apparently, Northampton was still:

*'important as the capital town of one of finest grazing and hunting counties.'*

So, not all bad then. Peterborough, however, comes in for more of a battering. Here, he says, is a city:

*'without population, without manufactures, without trade, without a good inn – or even a copy of the Times; a city which would have gone on slumbering to the present hour without a go-ahead principle of any kind.'*

One presumes there was also a dearth of small animals to pursue in this 'peculiarly stupid city'. The current home of the Nene Valley heritage railway, with more than enough 'population', manufacturing industry in abundance and plenty of good inns, Sidney might like it better now. He might even be able to get that elusive copy of the Times. Sidney returned to Blisworth on the London to Birmingham main line to continue his journey north, passing through the Kilsby tunnel – at 2433 yards long, once one of the wonders of the Victorian world – but by Sidney's day…

*'reduced to the level of any other dark hole.'* [98]

His destination was Rugby. The town then lay a mile from the station but is never otherwise mentioned. Sidney instead devotes several pages to eulogising about the public school and in particular their progressive headmaster Dr. Arnold. In consequence, in the interest of, well, 'interest' we shall therefore move on quickly to Coventry, which was his next port of call.

The Coventry that Sidney saw disappeared during the Second World War but some of the buildings he describes (St. Michael's Church, St Mary's Hall and the Free School – now the King Henry VIII School) are still there. Sidney complains about the 'curious specimens of domestic architecture' forming this 'dark dirty inconvenient city' so perhaps he would prefer its modern equivalent – or perhaps not. Coventry was then the manufacturing centre for 'ribands' (sic) and 'watches' and small enough to be 'surveyed… in a few hours' [99]. With seemingly little to detain him, Sidney swiftly moved on to Birmingham.

New Street station had yet to be completed and Sidney stepped from the train at Newton Road in the centre of 'a healthy ugly town' in a 'squalid' part of the city known as Hardware Village. Newton Road station has had a chequered history. The station where Sidney stepped off the train was built in 1837 by the Grand Junction Railway but was replaced, within fifteen years of Sidney's visit, by a new station just a short walk down the track at Ray Hall Lane [100]. Birmingham was then the centre of the national railway network – a meeting place for all major railway companies of the day, including Brunel's seven foot gauge Great Western.

Birmingham was also England's manufacturing heart and the air was black with soot from factory chimneys, so it's no wonder Sidney describes it as a 'very dull dark, dreary town' from which the reader is advised 'the sooner he gets out of it the better'[101]. He was nevertheless pleased with the nearby Station Hotel where there was a 'capital table d'hôte provided four times a day …. with servants included' for two shillings a head. Prices are assumed to have risen since then while it is also assumed servants are no longer included. Considering Sidney was such an advocate of free enterprise and industry, he seemed less enamoured of the nouveau-riche such enterprise generated. As may be gathered from earlier comments Sidney was a snob. The typical 'Brummagen Jupiter' (sic) is described as:

*'lion faced, hairy, bearded, deep mouthed swaggering, fluent in frank nonsense and bullying clap-trap – loved by the mob for his strength and by the middle classes for his money.'*

The business of Birmingham was, as today, conducted in the city centre but the less well-off lived in the suburbs 'in long streets crossing each other at right angles' where 'the houses are built of brick toned down to a grimy red by smoke'; the well-to-do had migrated to the surrounding country and purchased estates far away from 'the bickering cliques' of the middle classes. Button making was

Birmingham's main industry then, although it was also the centre for glass making and, more interestingly, the production of black lacquered, tin plated, papier-mâché filled articles of furniture such as fire screens. Sidney goes into some detail on the many types of button produced. It must have been riveting ('scuse pun) work since one woman alone 'could punch out 57000' button shells a day. The shells were welded together in a press worked by children aged seven to ten years – an incidence apparently so common that Sidney makes no adverse comment about it, although he does point out that kids working long hours for 'sixpence a day' are unlikely to get an education being 'only fit to sleep' by the time they finished their shift. He therefore suggests that factory owners make sure their employees are at least thirteen years old and capable of being able to read and write.

Sidney also visited a gun factory, a day out he commends to the reader, even though he concedes that 'the employment is hard and dangerous' since the metal grinding stones often break while in motion:

*'in which case pieces of stone weighing a ton have been known to fly through the roof of the shop; unwholesome because the sand and steel dust fill eyes, mouth and lungs, unless a certain simple precaution is taken which grinders never take'.*

What this simple precaution may have been is not divulged but dust masks would certainly not have been provided. Quill pens of the sort used by Scrooge in 'A Christmas Carol' were also produced in bulk in the city and their manufacture is described in monotonous detail, as is brass making, tool manufacture and electroplating. Whilst he applauds the industrious nature of Brummies, he deplores that despite their:

*'excellent …. Intelligent and ingenious qualities the people of Birmingham are much more dirty, drunken and uneducated than they ought to be'*

A consequence no doubt of the '1293 hotels, taverns, gin shops and beer shops' that supplied intoxicating liquor to the locals. In 1849, 1700 Brummies were taken into custody for offences arising from intoxication – and that was without any input from Aston Villa fans.

Before departing from Birmingham, Sidney makes one last comment about the generally poor state of health of its residents, and in particular the working classes pointing out that although:

*'a private water (supply) company exists... (it) has scarcely been called upon to supply the houses of the working classes... the filth in which fifty thousand people live seems only to be understood by the Medical Inspectors, whose reports have produced so little effect.'*

He concludes that it is therefore:

*'not extraordinary that after long hours of toil the inhabitants fly to the bright saloons of gin shops and the snug tap rooms of beer shops.'*

And it is among the bright saloons and gin shops of Birmingham we now leave him, his journey still only half completed, as he makes ready to benefit us with his opinion on the industrial heartland of Leeds and Manchester, taking in as much fox hunting as available along the way. Sidney is a difficult man to categorise. Today he would be seen as just another patronising upper middle class bigot yet, by the standards of the day, he was probably enlightened. If he was repulsed by the dirt and squalor in which working class people lived, he could see nevertheless that their miserable situation was beyond their control and arose from the inaction of the educated rich. Also to his credit, he had even less time for his own people, the middle class, who had foresworn the trappings of gentility in pursuit of mammon.

So is there anything to learn from 'Rides on Railways'? Well, what is most surprising is not the difference in towns and cities between 1850 and today but their similarities. The coming of the railway changed them for better and worse and, by the time Sidney saw them, they were well on their way to becoming the amorphous urban landscapes around us today. As Sidney noted, we live with both the benefits and consequences of railway mania.

**Notes to Chapter 10**

90. The connection to the Great Exhibition is apt as Sidney later became one of the Exhibition's administrators after it transferred to Sydenham.
91. The actual title of the book is in inverse proportion to the length of the book itself, being ' Rides on Railways leading to the Lake and Mountain Districts of Cumberland, North Wales and the Dales of Derbyshire; with a glance at Oxford, Birmingham, Liverpool, Manchester, and other Manufacturing Towns, by Samuel Sidney author of 'Railways and Agriculture', 'Australian Handbook' etc... Presumably with a title like this there would have been no room for a cover picture. Despite the book's obvious commitment to the London and North Western Railway, there is no evidence that Sidney benefited directly from the free publicity he gave the company.
92. He also wrote numerous books on Australia, a country he never visited, based on third hand information provided by his brother who lived in New South Wales.
93. An (usually) early morning train, resulting from an Act of Parliament that required railway operators to provide at least one cheap alternative to their normal trains.
94. As described in 'Dombey and Son' published 1848.
95. The sheds occupied an area of 135000 square feet, the goods platforms 30000 square feet with 110 cranes for loading and unloading.
96. Sidney was lucky to see the Camden Roundhouse being used for its intended purpose. Built by George Stephenson and capable of housing 24 locomotives, it had a short life as an engine shed as it proved incapable of accommodating anything other than the shortest of locomotives. It was sold off in 1869 and became a liquor warehouse for the next fifty years, following which it was abandoned until the 1960s.
97. The Great Western Station (still used today) opened two years later in 1852.
98. The workshop and foundry were used solely for repairs until 1845.
99. It's a pity Sidney was so sniffy about the Kilsby Tunnel; here was one feat of engineering that one could be justly proud of. Trial shafts drilled along the proposed line had failed to reveal the presence of quicksand. Robert Stephenson, who was the engineer, tried pumping away nearby springs but to no avail as one day the roof collapsed and the bricklayers, who were lining the roof, nearly lost their lives as flood waters engulfed the dig. The waters were eventually overcome by pumping at the rate of 1800 gallons/minute for eight months.

100. The city was surrounded by 'Lammas lands' which were restricted for use only to city freemen for the purpose of animal grazing.
101. Ray Hall Lane Station lasted forty years. It was replaced with another station at Newton Road not far from the original site. The site of the former Ray Hall Lane station is now somewhere under the M5 Motorway.
102. Travellers were apparently always male in Sidney's day.

# Chapter 11

## *Christopher Tennant, railway entrepreneur*

*Portrait of Christopher Tennant*

When George Stephenson died in August 1848 his *'remains were followed to the grave by a large body of work-people'*. Businesses in Chesterfield, where he lived, closed for the day as a mark of respect and the funeral procession was preceded by the *'corporation of the town'*. His importance to his local community and the nation in general was rightly being recognised. A similar show of respect and affection was accorded in the passing of Christopher Tennant. At the time he died he was living with his mother on the headland at Hartlepool and the townspeople turned out in force to show their respects. Sadly, in contrast to Stephenson, his name is now no longer remembered. Even in his adopted town his grave lies forgotten and the headstone broken. The harsh sea air has eroded the inscription on the stone but just about discernible are the words:

*'He was distinguished by his perseverance and patriotic exertions in promoting various public works… to the prosperity of the country'.*

The 'public works' referred to were the development of the docks and railways in Hartlepool and Stockton-on-Tees. Tennant was one of the world's first railway entrepreneurs and the eulogies confirmed on him in his passing show he was considered by contemporaries as a major player in the development of railways, so who was he and what indeed was his contribution?

He was born in 1781 in the Yorkshire market town of Yarm, where his father owned a hat making business. It is said, without any real evidence, that he spent some time in the Royal Navy before moving to nearby Stockton, where his brother ran a sail and rope-making company. His move to Stockton coincided with the most significant period in the town's history. While his naval contacts would have proved useful to his brother's interests, Tennant's ambitions lay elsewhere. Never wealthy, he somehow raised enough money to purchase lime kilns at Thickley near Shildon. Shildon and Stockton are names now synonymous with the Stockton and Darlington Railway yet the direct route from Shildon to Stockton on the Tees was not via Darlington, as the Pease family proposed. This necessitated an unnecessary dogleg to accommodate the connection to the Quakers' town.

On behalf of Stockton business therefore, and using his own money, Tennant, in 1818, commissioned a survey for a canal that took the straight line route to the coast. This conspicuously avoided Darlington, the estimated cost being £205,000 and, inevitably, the route chosen met with vigorous opposition from the Pease family. An enmity thus developed which would confound Tennant's lofty ambitions for the rest of his life. The upside of Tennant's survey was that it generated a lot of commercial interest. This, in turn, encouraged the Pease family to find a cheaper alternative to the construction of a canal, which had also been their first thought. Instead, they considered, for the first time, the idea of a railway. Their railway would follow roughly the original line chosen for their canal scheme, thereby reinstating the Darlington dogleg. They duly commissioned another survey, but this time for an 'iron railway'[102]. Three more years would pass by before the railway scheme gained royal assent, at which time point we reintroduce George Stephenson.

George is widely considered to be the first true railway pioneer. Never the genius that, say, Richard Trevithick was but an enthusiastic self-taught engineer with a consuming passion for steam locomotives. Until George's appearance, there were no plans to run locos on the Stockton and Darlington Railway (S&DR) as it was intended to be a simple mineral freight operation using horse-drawn wagons.[103] Inspired by George's enthusiasm, the new railway was rejigged to a specification which, for the first time, included a commitment to steam locomotives. It opened for business in September 1825.

Ironically, as it would prove, one of the first customers of the Stockton and Darlington Railway was Christopher Tennant who needed limestone and coal delivered to his kilns at Thickley and the manufactured lime transported to Stockton. Bearing in mind his earlier canal proposal, it can only have annoyed Tennant that coal, lime and other goods now had to undertake an unnecessary eight mile detour for no other reason but to appease the Peases. The crunch came when the S&DR unveiled plans to extend their railway nearer to the mouth of the Tees, where the river was deeper and easier to navigate, but, tellingly, beyond Stockton to what became the town of Middlesbrough. Whilst making economic sense, it effectively side-lined Stockton as a port. Tennant, who was a Stockton resident and local leading establishment figure, decided to take matters into his own hands.

In anticipation of this, months before 'Locomotion' steamed into Stockton on that fateful day, in 1825 Tennant was already planning another railway. This would connect the still isolated South West Durham, Weardale, coalfields at Willington to new docks he was going to build at Haverton Hill, to the east of Stockton. The idea wasn't new, being a revival of an earlier proposal for a 'Tees and Weardale Railway', which was rejected by Parliament in 1819. Tennant changed the original proposal, making it more palatable to investors by adding additional branch lines to other Durham collieries bypassed by the S&DR. He was still on speaking terms at the time with the Pease family and offered, in the first instance, to build the new railway jointly with them, albeit employing his new port of Haverton Hill as preferred terminal. The S&DR turned him down flat and he was forced to strike out on his own.

In theory, prospects for his railway looked good but he was opposed by a combined force of Tyneside coal merchants and landed gentry; in particular, Lord Londonderry through whose Wynyard land the railway was meant to pass. As a result the S&DR succeeded in getting Tennant's Parliamentary Bill thrown out. However, he was not easily dissuaded. He revised the proposal yet again, this time shortening the line, so that instead of running all the way into Weardale it terminated just a few miles beyond Darlington in the very heart of S&DR territory, incidentally providing the more direct route to the sea he had always sought.

*Stockton and Hartlepool Railway train 1841*

The new railway, linking what became Port Clarence, on the north bank of the Tees, to the S&DR line at Simpasture Farm (now part of Newton Aycliffe) was to be known as the Clarence Railway (CR), the name a sop to the current Duke of Clarence and future William IV. In this guise, it received Parliamentary approval on the 23rd May 1828. Tennant was the main shareholder in the company, the only instance of his ever holding shares in any of the companies he instigated, promoted or developed.

Unfortunately, this upstart was now in the position of siphoning trade away from the S&DR and their directors were apoplectic; 'War... open or concealed', according to Messrs Pease, had been declared.

And war it would be; a war that, in their current situation, the Clarence Railway had no hope of winning. The problem was that the CR needed to use a section of the S&DR rails beyond a junction near Simpasture Farm, near Shildon, and from the outset the Quakers set out to make things as difficult as possible for their unwanted tenant.

A situation which began badly deteriorated rapidly. Tennant needed finance from the main financial institutions of the day for his railway and the 'Society of Friends' closed ranks to deny him access. This seriously delayed the completion of the CR and the moment it opened for business the S&DR increased their charges on the leased stretch of their railway. Additionally, they insisted that CR wagons must go over a weighbridge before being allowed to travel over their short stretch of line. Unsurprisingly, their own wagons were just counted past a checker. This had the anticipated effect of increasing the time taken for CR freight to move to the docks, thereby obviating the shorter distance to the sea involved on the CR compared with the S&DR. Additionally, the S&DR refused 'for safety reasons' to allow CR horse-drawn wagons the use of their rails outside daylight hours, although no such restrictions applied their own horses. The effect of all this on CR finances was disastrous. The result was that the Clarence Railway only began to pay a shareholder dividend after they built their own line to the central Durham coalfields via a branch at Byers Green.

Tennant, by then, had moved away from Stockton to join his mother at Middlegate in Hartlepool where he displayed as much enthusiasm in promoting the interests of Hartlepool as he had for Stockton. He immediately threw his weight behind a previously abandoned project, dating from 1823, to construct a new dock in Hartlepool's old and hitherto neglected natural harbour.

As noted, Sir John Rennie was commissioned by the dock and railway committee to produce plans for what eventually became Victoria Dock, with the construction of the connecting railway handed to George Stephenson. Acting as overseer for all these projects was Christopher Tennant, having been made General Superintendent of Works of the Hartlepool Dock and Railway Company (HD&R) on a salary of £400/annum. The fact that he was an employee of the HD&R and not a shareholder or director gives some indication as to his financial situation. Because of his first-hand knowledge of Parliamentary procedure, his first official duty was to go to London and steer the Company's Railway Bill through; the completion of which, in 1832, earned Tennant a vote of thanks from the HD&R Committee. The successful construction and rapid growth of the HD&R is detailed elsewhere, but it is to Tennant's credit, given his limited engineering experience, that the works proceeded fairly smoothly to completion.

*Middlegate, Hartlepool*

From his involvement with the HD&R, Tennant saw the benefit of linking his own Clarence Railway to the town where he now lived, via another railway to the docks at Hartlepool from the south, effectively creating a further export outlet for South Durham coal, albeit it again in direct competition with the S&DR. Consequently, he threw his weight behind what became the Stockton and Hartlepool Railway (S&HR) and the solicitor who acted for Tennant on this was Ralph Ward Jackson, who went on to become the founding father of the new town of West Hartlepool.

*Former Stockton and Hartlepool Railway, Mainsforth Terrace, Hartlepool*

What sort of haulage media did Tennant employ? Although he continued to use horse-drawn wagons, he soon assigned the bulk of the work to steam locomotives. His old enemies, the Pease family, had a significant stake in Robert Stephenson's engine works at Newcastle and would no doubt have been reluctant to assist their main opposition in any way. Additionally, Timothy Hackworth, the one other noted local locomotive builder of the time, was directly employed by the S&DR until 1833 and was expressly forbidden to work on engines for other railways[104]. As a consequence of this, the first engines were almost certainly supplied by Edward Bury who also provided the first H&DR engines. In later years, both the Clarence and Stockton and Hartlepool Railways purchased engines from 'Hackworth and Downing' at Shildon and, as we know, Thomas Richardson of Castle Eden.

Tennant's battles with the S&DR continued to the end of his life. In the 1830s the S&DR made a bid for the lucrative main north to south rail route between London and Edinburgh to operate over their rails. Tennant, quite reasonably, saw no reason why the route shouldn't hug the coast, which in terms of existing population centres and industry at the time made more sense. In this, he was again stymied by the combined political and financial machinations of the Quakers of Darlington. And it was therefore Darlington, not Hartlepool or

Stockton, that became the unlikely North East focal point for all things rail for the next hundred and thirty years.

Tennant died in 1839 whilst working in Leeds trying to negotiate the establishment of the UK's first 'fish' train to that city from his home town of Hartlepool. He never married. He is buried in his mother's grave at St. Hilda's church on the Headland close to where he once lived.[105] The memory of Christopher Tennant, both the man and his achievements, has long faded. In his birthplace, Yarm, there is no evidence of his existence and the same is true in Stockton. Hartlepool, perhaps more surprisingly, also has nothing to offer by way of tribute, despite the fact that Tennant was a major factor in the growth and wealth of the town.

If Tennant left us a legacy, therefore, it is in the continuing importance of the current east coast rail network north of Teesside and in the establishment of Hartlepool and its associated dockland (and to a lesser extent that of West Hartlepool). For better or worse, the vast industrial complexes to the north of the Tees, such as those belonging to I.C.I. who once employed nearly 10000 people at their Billingham site alone, would have been impossible were it not for the rail links Tennant created. We can only imagine, how different the landscape of the North East would have looked if the Pease family had achieved the rail monopoly they sought. Unlike other railway entrepreneurs of the time, Tennant never conspicuously sought wealth or fame for himself. His reward was what he achieved for his fellow man and, for this reason alone, I raise my hat to him.

**Notes to Chapter 11**

103. Tennant's desire for a direct route from Stockton to the coalfields also extended to compromising with the S&DR in proposing a direct railway with separate branches to Yarm and Darlington. His idea was rejected by the S&DR board.
104. Indeed the corporate seal shows a horse pulling a line of coal wagons.
105. He was given special dispensation to build 'Sanspareil' which took part in the Rainhill trials.
106. Interestingly, the church is less than 50 yards from the committee meeting place (the Kings Head Inn) of the 'Great North of England, Clarence and Hartlepool Railway' one of Tennant's railway companies' rivals in later years.

# Chapter 12

## *Aled Roberts' life in railways*

*Aled Roberts in the cab of former GWR 75xx class locomotive no. 7754 at Llangollen*

Aled Roberts had just clocked on for the early shift when he was informed about the accident. He remembered 'a terrible silence' in the yard. The charge hand's first words to him were, "Aled, the 'Mail's in the river at Llangollen."

The Second World War had just ended but in that summer of 1945 the weather didn't reflect the general euphoria prevalent throughout the land. Typical of this was a terrible thunderstorm in Flintshire that left behind an aftermath of floods and the River Dee was threatening to burst its banks. However for a few days the weather improved and by the early hours of the 7th September, when the 3.35 mail train left Chester for Barmouth, the weather was fine, dry and clear[106]. To the west of the town of Trevor, where the railway entered the Dee Valley, the rails had been laid on a narrow terrace, forty feet below the Shropshire Union Canal. The railway here formed part of the Great Western (GWR) network but the canal above it was in the ownership of their rivals the London Midland and Scottish (LMS). At around 3.30 am, for reasons that remain unclear, the canal burst its bank and the ensuing torrent swept away 120 feet of railway embankment, leaving just the rails hanging in space above a water scoured crater. Given the hour and remote location (2 miles from Llangollen and 2 miles from Trevor) there were no witnesses and, perhaps worse, the telegraph lines parallel to the track were unaffected so no 'line fault' warning was transmitted to the signalmen on each side of the landslip[107]. Accordingly, with signals at 'clear', the mail train approached, gathering pace on the downhill slope to Llangollen.

*The Llangollen rail disaster of 1945*

On the footplate of GWR '4300' class engine 2-6-0 number '6315' were two of Aled's workmates Driver Dai Jones and Fireman Geoff Joy, and in the guards van was another of his comrades, Fred Edwards. The mail train hit the breach in the embankment at 4.51am and the locomotive left the tracks, pitching vertically down the steep hillside into the river, with the coaches bunched up behind. The wreckage caught fire and since it would be another hour before help arrived little survived the ensuing inferno.

Locomotive 6315 fell nose first into deep mud and driver Dai Jones was crushed in the cab and killed instantly. It would seem he spotted the subsidence before the train plunged into the river because he was found with his hand tightly gripping the brake lever. Fireman Joy was thrown clear but broke his wrist in the fall. However he still managed to scramble up the hill side, through a cascade of mud slurry, to the Wrexham to Llangollen road and ran the remaining 1½ miles to Llangollen to raise the alarm. Fred Edwards meanwhile had left the guards van which, unlike the rest of the train, was undamaged but suspended over the precipice and hurried in the opposite direction to alert the signalman at Trevor. Amazingly, Jones and Joy were the only casualties of an otherwise empty train.

Aled was a witness in the subsequent enquiry. When he arrived on the scene the following day what remained of the mail train was still burning. The locomotive, though not badly damaged proved impossible to recover and had to be cut up on the spot. The enquiry reached no conclusion as to what caused the breach in the canal wall but spring activity beneath the canal and poor maintenance of the retaining embankment were cited as likely suspects.

Fortunately, most of Aled's long life on the railways was neither as dramatic nor as sad. When I met him in 2001, Gwilym Aled Roberts, to use his full name, had recently completed his sixtieth year on the railways. A genial Welshman, and the son and grandson of railwaymen, he was then acting as a volunteer on the Llangollen heritage railway, being the mentor to trainee drivers. As a local he knew both the railway and the surrounding mountainous landscape well. He came from Trawsfynedd, a village north of Dollgellau, later famous as the insensitive location of a nuclear power station. Virtually all his working life was spent in the area on the former GWR lines south of Liverpool and north of Birmingham.

He started work, aged fifteen, as an engine cleaner at Crewe at the height of the Second World War but was soon transferred to Wolverhampton. With Birmingham the target for Hitler's bombers Aled was employed night and day keeping Department of War Transport trains running. He transferred to the motive power depot at Wrexham before the war ended and was there for the rest of his working life. Gradually he worked his way up the arcane layers of railway apprenticeship to fireman and, finally, engine driver.

*Site of former Arenig station*

As a driver he was particularly at home on the lines that weaved their tortuous way through the Welsh mountains. Welsh was his first language so one immediate advantage he had was in being able to pronounce tongue twisting place names such as Glyndyfrdwy.[108] The railways in this part of the United Kingdom had to cope with the worst that British weather could offer and the line north of Bala to his birthplace at Trawsfynedd was particularly vulnerable. It ran between the remote mountains of Arenig Fach and Arenig Fawr, ascending a steep valley through a small hamlet called Capel Celyn. It was in this wild place that a GWR '22XX' class engine Aled was driving once came to a halt, wheels spinning furiously in a howling gale and driving rain. The locomotive was fronting a 12 coach troop train packed with 400 soldiers, heading for the camp and gun ranges near his home village and the reason the engine stalled was that the locomotive's sandbox had run dry and the wheels were therefore finding difficulty gaining traction. Aled inched the train forward to Arenig where he uncoupled the engine. He then took the loco on to Trawsfynedd, refilled the sandbox and returned to rescue the unheated train. The troop train jam packed with shivering squaddies eventually arrived at their destination 3 hours late. There is an aside, if unrelated directly, to this story. Aled's wife is from the nearby village of Capel Celyn, although they met for the first time when Aled was living in Wrexham, and the former hamlet of Capel Celyn now lies, along with the track bed where the engine stalled, at the bottom of Llyn Celyn reservoir. [109]

Aled's local knowledge of the area and close familiarity with the extreme mountain weather conditions meant he was always favourite to get dragged out of bed to work the snow ploughs from the town of Bala, usually astride the footplate of the Chester based engine '2313'. Bala at the time, amazingly for its size, had three stations; Bala Central, Bala Junction and a halt at Bala Lake. Bala Junction was located at the junction of the Ruabon to Barmouth line, from where there was a northern spur to Blaenau Festiniog. When the line to Blaenau closed in 1961 the short section from Bala junction to Bala was briefly retained[110]. In consequence the main station, Bala Central, now stood in splendid isolation, linked only to the rest of the rail network by a single coach shuttle service to Bala Junction. Nevertheless, regardless of paucity and variety of trains, the Station Master at Bala Central, Tom Green, took his duties seriously. Before each departure he always made a formal announcement over the station public address along the lines,

*'The next train from Platform 1 is for Bala Junction. Change at Bala Junction for Liverpool, Carlisle, Glasgow and Birmingham for trains to London etc.'*

He extended and varied the announcement to suit his mood and potential audience, changing the number and distance of potential destinations accordingly. Bala specialised in 'characters'. One of the porters was always known as Dai 'Tatws'. Tatws is Welsh for potatoes and Dai had a side-line in potato sales which were sold for beer money to passengers on the station platform. During breaks between shuttle services the train crew would adjourn to the nearest pub. On one such occasion they overstayed their welcome and only made it back to the station immediately prior to the train's departure. They leaped aboard and started away immediately. When they arrived at Bala Junction they found they had left the train's single coach, along with its disgruntled passengers, behind at Bala Central.

The line from Ruabon to Barmouth ran through some of the most spectacular country in Wales. Following the Dee Valley from Llangollen to Lake Bala it descended the valley east of the mountain of Cadair Idris to Dolgellau where it then skirted the southern edge of the Mawddach estuary before joining the Cambrian coast line at Barmouth. There was a tiny engine shed at Penmaenpool that Aled sometimes worked from. It had only three or four engines, all dedicated to the few miles of line between Dolgellau and Barmouth. Next to Penamenpool, as today, there is the George III public house and the landlord, being a personal friend of the railwaymen, often put out bottles of beer on a trackside post for the drivers to pick up as their engine passed by.

Dolgellau, Bala and Llangollen were busy stations then but other stations along the line, such as Berwyn, were barely used by the travelling public. Despite this Berwyn retained a full quota of staff, to whit three signalmen, three porters and a station master who were only there to serve the needs of the nearby 'Big House', whose occupier permitted wayleave rights over his land. Aled remembers 'Bob' the station master as an obnoxious character with an inflated view of his own importance. He would run out on to the platform the moment the train pulled to a stop ordering the engine crew to load coal buckets from the engine tender for the station master office. This they reluctantly agreed to apart for one particular wild and windy night. Tired and running late, they filled Bob's coal bucket with water, ensuring that the first application of coal would give him with a taster of footplate life in foul weather.

Aled retired in 1991, having completed his fiftieth year for the GWR, British Railways and successor organisations. Almost immediately he was roped in to drive the engines on the Llangollen Railway, a line that follows his former Ruabon to Barmouth line from Llangollen to the station at Carrog[6]. Since he had driven most types of 'Western' locomotives including those from the engine shed at Llangollen, such as 'Foxcote Manor', he was constantly in demand, particularly if a TV or film company needed an experienced engine driver for a locomotive in a period drama or documentary. Even when he considered himself too old to drive his skills were still employed as footplate inspector. At the time this piece was written he was still a regular visitor at Llangollen, chewing the fat with the staff there. Of the lines he worked near his birthplace little remains. The former track-bed north of Bala, through Arrenig, can still be seen winding its way along the side of the mountain above the road to Trawsfynedd, although, as mentioned, it disappears for a time beneath the still waters of Llyn Celyn. Sections of the line to Barmouth between Llangollen and Carrog and Bala to Llanuwchllyn have become 'heritage' railways although the latter is now narrow gauge. The station at Corwen is still there, albeit converted to office premises but no trace remains of the once busy station at Dolgellau (buried beneath the Dolgellau bypass). The track from Penmaenpool to Barmouth is now a popular cycle way and footpath.

*Winter at Carrog on the Langollen Railway*

---

6   Now extended to Corwen

**Notes to Chapter 12**

107. 'Report on the Accident at Sun Bank Halt Llangollen' by Lieutenant Colonel G.R.S Wilson on behalf of the Ministry of War Transport.
108. In fact the first hint that something drastic had happened was the loss of water supply to the Monsanto Chemical Works at Chirk. as the works drew their water from the Llangollen Canal.
109. The stone quarry is now disused but the former marshalling yard forms part of a car park providing access to walkers to the Arrenig mountains.
110. A station 1 mile east of Carrog,; it should be pronounced, as Aled does, 'Glin – dove- er- doo- ee' but tends to get shortened by staff on the Llangollen railway to the easier (for English visitors at least) 'Glin–dove-ree'.
111. A short section from Blaenau to Trawsyfnydd did reopen in 1964 to serve the nuclear power station.

# Chapter 13

## *Isombard Brunel and the 'Battle of Mickleton Tunnel'*

*Isombard Kingdom Brunel*

History has been kind to Isombard Kingdom Brunel. In turn revered and reviled in his lifetime, he is now considered one of, if not *the* greatest, engineer of the Victorian age, to the point of being nominated in the top ten Britons of all time, albeit by Jeremy Clarkson. There have been so many biographies of the man one would suppose there is little to add, however one incident bears recounting since it casts a different light on his complex character. It occurred because of his involvement with the Oxford Worcester and Wolverhampton (OWW) railway, which arrived at a time in Brunel's life when he could well have done without the hassle.

1845 had been a bad year for Brunel. His father and mentor Marc suffered a stroke that left him paralysed for the rest of his life and Isombard, whose time was spread as thin as the hair on his receding pate [111], had recently thrown his weight behind the expensive and ultimately embarrassing enterprise that was the building of an atmospheric railway,[112] on the South Devon Railway (SDR). Brunel, in his own words, was *'not an advocate of this or any other system'* nevertheless he had been lured by the idea that trains might operate without the need for (expensive) locomotives. Moreover, experiments had shown that atmospheric trains could potentially work at higher speeds and over steeper gradients than was currently possible using steam traction. He even conducted his own trials on an experimental length of track at Wormwood Scrubbs and on the one operational atmospheric railway at Dalkey in Ireland, becoming convinced that this novel 'system' could be made to work. Sadly it couldn't, at least not with the technology available.[113] Too late he discovered that the;

- pumping engines used to maintain the vacuum in the pipe work kept malfunctioning.[114] It only took one to fail for the whole railway to shut down.
- pipework manufacturers had difficulty casting the continuous lengths of pipe required.
- junction between the complex valves on the unit linking the lead coach and the vacuum pipe had recurrent sealing issues.

So while, in the summer of 1845, Brunel was still producing optimistic reports on the progress of his atmospheric railway inside he knew he was in deep trouble. Nevertheless, being Brunel, he persevered.[115] So, with all these problems on his balding pate, did he ever get involved with the OWW?

The OWW (later to be nicknamed the 'Old Worse and Worse') had been created the previous year in a meeting fronted by the Mayor of Worcester in Worcester Guildhall[116]. The motivation behind it was an attempt by businessmen to break the monopoly of the 'London & Birmingham Railway' on goods movements from the Midlands to London and the South East by establishing a route through the Cotswolds to the Great Western Railway's (GWR) rails at Oxford. As a consequence, the GWR were keen to be involved, with five of their directors assigned to the board of the new company that would achieve this goal. The GWR even agreed to guarantee a 3.5% annual dividend to OWW shareholders in exchange for 999 year running rights over the line. Their interest was understandable given that, at this time, their 7 foot 'broad gauge' was being denied access to the industrial heartland of Birmingham. They must therefore have seen the OWW as a means of entering the Midlands by the back door. Additionally, they could also see that, by joining forces with the OWW, they could fend off those standard gauge interlopers trying to elbow their way into traditional GWR country from the north – standard gauge, for example, was already at Gloucester.

The OWW was given parliamentary sanction on the 4th August 1845, with a clause in the Act which allowed the GWR to step in and complete the railway should the OWW default in its construction. [117] However, in the same paragraph, and controversially, the GWR under the same circumstances could actually be *made* to complete the work, if required to do so by the Board of Trade – an issue that would return to bite them later. We can assume therefore it was at the insistence of the GWR that Brunel was nominated official Engineer of the OWW.

The building contract was given to a company called 'Peto and Betts', who had worked previously for the GWR with Brunel overseeing the work. This was underfunded and beset with financial problems from the start.[118] The route chosen was relatively straightforward but included a half mile long tunnel at Mickleton, north of Chipping Campden. On paper, it was a straightforward engineering operation but one which would add considerably to Brunel's woes over the coming years. The subcontractor given the task of building the tunnel was a company called 'Gale and Warden' (G&W), who withdrew after 18 months, complaining of technical difficulties (including running quicksand). Their problems, both practical and financial, were exacerbated by an unnecessary three month delay by the OWW in purchasing land which prevented them working during the productive summer months. The situation worsened with the appointment of their successor, 'Ackroyd, Price and Williams' (AP&W). History has defined AP&W as the villains in the ensuing pantomime but the vast bundles of litigation documents later compiled by both sets of lawyers suggests the resulting chaos was as much the fault of employer as employee.

The first salvos in the conflict were issued from the OWW camp. The OWW claimed that tunnel construction had fallen way behind schedule (*which it had*), that the quality of the work completed was poor (*which it was*) and that AP&W were dragging their feet in order to pressurise their employer into providing additional funding to guarantee the tunnel's completion on time (*more than likely*). The contractor in turn argued that they hadn't been paid for work completed (*probably true*) and needed the money desperately to finance the next phase of the work (*also probably true*). Claim led to counter claim, and tunnel construction progressed at Thomas the Tank Engine pace for another three years, by which time Ackroyd and Price had departed the now Williams and Co. to be replaced by Robert Mudge Marchant.

Marchant is a fascinating character. He had worked for Brunel on other railway projects and it must be assumed he was Brunel's choice as Resident Engineer to supervise the tunnel's construction. In this there is little evidence to suggest he did not carry out his employer's instructions to the best of his ability. The trouble was he was in the awkward position of playing referee in a contest between two equally disreputable opponents. The litigation file is filled with letters from Marchant to Brunel complaining about the poor quality of the work and the low calibre of men employed, with similar, if more belligerent, correspondence from Marchant to Williams[119]. Whilst he was conducting this battle he was simultaneously having to act as apologist for the OWW whenever the contractor didn't get paid, which was often. In other words Marchant couldn't win.

What motive Marchant had for throwing in his lot with Williams & Co. we'll never know but after doing so he continued to act honourably, as far as is possible to determine, putting in £10k of his own money to upgrade the ancient equipment used at the construction site and recruiting a more skilled workforce. Sadly, the

OWW had by then decided it wasn't prepared to release another farthing until they received an independent evaluation of what had already been achieved. Construction work came to a halt and the stand-off between employer and contractor continued for a further two years. Eventually, an exasperated Brunel gave Williams and Co. an ultimatum; if they didn't pick up their shovels again he'd step in, as indeed required by the Parliamentary legislation, and complete the job himself. This was in March 1851 and Marchant, by that time, was claiming he was owed more than £4000 by the OWW and no longer able to pay the wages of his own staff. Marchant delivered his own ultimatum; the OWW must pay him at least a £1000/month before he would allow his men to carry out any more work. To Brunel's credit, he saw merit in Marchant's claim and initially backed the contractor against his employer. Unfortunately, the OWW didn't. They invited the main railway contractors Peto and Betts to step in and take possession of the tunnel workings and equipment and, in the process, fling Marchant's merry men out on their ears.[120]

So began the Battle of Mickleton Tunnel.

*GWR 'Castle' class engine leaving Mickleton Tunnel*

Early skirmishes took place in June when Parson and Peto sent in a host of heavies to take possession of the site only to be soundly beaten by Marchant's men who were dug in. This was the signal for Isombard Kingdom Brunel to enter the contest.

Whatever he felt regarding the rights or wrongs, Brunel was first and foremost a company man and the project, his project, couldn't be allowed to founder on the whims of a few disgruntled diggers. He recruited a couple of hundred 'Peto' men, and armed with pickaxes and shovels (and oiled with gin), they marched on Mickleton. On the 20th July they had their first encounter with Marchant's men who were manning hastily constructed ramparts at the tunnel's northern end. Marchant had known Brunel's army was coming and was well prepared. He initiated his own legal proceedings against the OWW to recover the money he claimed he was owed and forewarned the local constabulary that the OWW might try and force entry. In consequence, when Brunel turned up he not only had to deal with Marchant's mercenaries but also two local magistrates who accused him of 'disturbing the peace'. Brunel, who had volunteered as a special constable on occasions[121], was unwilling to break the law and withdrew his men. In doing so, however, he gambled that once things quietened down he could return and easily overcome the guards posted at the tunnel ends. Consequently, he turned up the following day but this time to his annoyance, in addition to encountering the same magistrates as before, he was confronted by policeman waving cutlasses. The magistrates read the Riot Act and Brunel again retreated, taking with him his booze bolstered battalion, who, having the calibre of today's football hooligans were 'up' for the fight.

Brunel used the next 36 hours to muster forces. He pulled in navvies from other OWW construction sites and from works at Warwick and Cheltenham and all were ordered to march on Mickleton. In the early hours of Monday the 23rd July one of the platoons passed noisily through Chipping Campden, where their carousing alerted the police who despatched a rider to Coventry to enlist the aid of the regular Army.

The first battalion of Brunel's barmy army, led by the man himself, arrived at Mickleton Tunnel at 3am where they met Mudge Marchant, regaled in cowboy outfit, with pistols sprouting from every orifice. Marchant warned Brunel he'd shoot the first man trying to get past him. Brunel considered this a second then ordered his men on to the attack. The navvies ignored Marchant and laid into his gallant but outnumbered band of followers. Marchant's defenders were overwhelmed and, despite Marchant's warning, in the end not a shot was fired. Brunel duly gained possession of the tunnel workings, with the only injury of any note that of a Marchant man who rashly brandished a pistol and was hit on the head with a shovel.

Round 1 to the short guy with the stove pipe hat and big cigar!

Marchant was not yet done however. He galloped off to alert his pet magistrates who in turn recruited thirty policemen and soldiers from the Gloucestershire Artillery, who arrived on the site within two hours.[122] The Riot Act was read again, with less impact this time, as uncontrolled punch-ups were breaking out right across the site. By now, the full force of Brunel's army had arrived and had gained control. One of the magistrates, convinced Marchant was legally in the right, unwisely advised him to order his men back to work. Since the tunnel site was now occupied by Brunel, this was like ordering the soldiers at Rorkes Drift to break out and engage the Zulu multitude in open combat. Nevertheless, they gave it their best shot and in the ensuing battle, which lasted a couple of hours, there were further casualties, including one man who had his features randomly rearranged with a shovel and another who had his finger bitten off. Luckily there were no fatalities[123]. With hand-to-hand struggles going on all around, Brunel desperately tried to restore order. Luckily, Marchant capitulated, conceding that the rights or wrongs of the matter were best decided in a court of law. The battle of Mickleton Tunnel wound to a close at the very moment the 'cavalry', in the guise of the regular army from Coventry, arrived on the scene.

Game, set and match to the vertically challenged Victorian. Well, not quite. The OWW replaced Williams and Co., as the tunnel contractors, with a company by the name of 'Hawley, Eales, Turner, Holden and Bryant', but were legally obliged to re-engage Marchant's men under the same terms as before. They were also made to come to a financial settlement with Marchant more or less along the lines he originally sought. Brunel never subsequently discussed the matter, so it is unclear what he personally thought about it, but Marchant, if not exactly a friend, was nevertheless a former work colleague and the Battle of Mickleton Tunnel, coupled with the OWW's eventual reluctance to employ his beloved 7 foot gauge, must have come as a huge blow to his pride. The following March he tendered his resignation to the OWW, offering to complete only those works in which he was directly involved. To this the directors of the company reluctantly agreed. By then the railway had been in existence for 8 years yet only four miles of it were operational.

*Mickleton tunnel today*

As Brunel rightly suspected the OWW never did use seven foot gauge engines on their railway[124]. Indeed, the OWW owned no locomotives or rolling stock of any kind by the time the railway opened fully for business later that year, having to acquire motive power at short notice from other railways. The construction costs had spiralled, to the point the OWW even tried to make the GWR pay for the line's completion, citing the Board of Trade clause in the original Act. Despite all this, the railway continued to expand, adding branch lines to serve local industry. These included an extension to Shrewsbury, via a controlling interest in what would one day become the heritage Severn Valley Railway; a part of the empire still worked by steam today. Shortly afterwards, the OWW itself was absorbed into the West Midland Railway, however by then Brunel had moved onwards and upwards. He had greater battles to fight over the following few years before his untimely death in 1859, though none were quite so contentious as the Battle of Mickleton Tunnel, where Brunel gained the unwanted reputation of being the last commander of a private army in a conflict that took place on English soil.

## Notes to Chapter 13

112. At this time Brunel was, in addition to the OWW, either acting or consulting engineer for the Great Western, the Bristol and Exeter, the Oxford and Rugby, the Bristol and Gloucester, The Newcastle and Berwick, the aforementioned South Devon etc. Not to mention his involvement in the construction of an engine works at Swindon, giving evidence to Parliament on other potential railways plus the Broad v. Narrow gauge controversy, the completion and maiden voyage of his iron ship the 'Great Britain' etc. etc.

113. Atmospheric Railways operated on the principle that rolling stock could be propelled by atmospheric pressure alone. The system chosen by Brunel was one employed at Dalkey, and later on the London and Croydon Railway, which involved evacuating a pipe laid between the rails by using strategically located pumping engines along the line. The leading coach was connected to the pipe via a complex rod and piston which opened a sealable valve allowing air into the pipe behind the valve; the air pressure then propelled the piston, and attached coaches, along the line of the pipe.

114. Robert Stephenson had pointed out in a report to Parliament that, amongst other obvious faults (such as being only able to operate in one direction i.e. no reverse function), it was totally inflexible, requiring complicated crossings and junctions whenever it encountered conventional railways. As Stephenson also presciently noted, any failure of the pumping engines anywhere on the system would shut the whole railway down.

115. Rolt suggests that Brunel's boiler pressure specification for the pumping engines was beyond the capabilities of current engineering.

116. That the motive power provided for the SDR opening day, the following year, was steam locomotives is a telling indication of the difficulties he was experiencing.

117. The shareholders names included one 'George Hudson' although this apparently was not the 'Railway King' himself, but given subsequent events, it might well have been.

118. Although the Act of Parliament specifically nominated a 7 ft gauge, the OWW laid down standard gauge rails and only eventually put in one 7ft gauge track down alongside the conventional rails when forced to do so. It was never used except in the early Board of Trade trials but dual gauge track was later laid, although rarely used, on other parts of the network.

119. Brunel suggested a total cost of 1.5 million pounds for the railway. This was then incorporated into the prospectus but Brunel was well aware this was a third less than the projected cost of other similar GWR lines

he had engineered. The OWW would subsequently have to go back to Parliament to raise more share capital for its completion. Why Brunel chose to under-price the job is not clear.

120. One bone of contention was the poor quality of the brickwork lining the tunnel. Brunel had specified the use of 'Roman Cement' as mortar but whatever it was Williams and Co. was using it wasn't Roman Cement and the roof of the tunnel kept collapsing. Not the best safety feature for a working railway.
121. A partner of 'Peto and Betts', by the name of John Parson, was now acting as paid legal advisor to the OWW; a lifelong opponent of the GWR, he later joined their board.
122. For example, during the Bristol riots of 1831.
123. There is no evidence that the police took any active part in the ensuing battle.
124. It is ironic therefore that, on the official opening day (17/5/1853), the train loaded with officials mowed down and killed a navvy stationed with a lamp near the tunnel entrance.
125. Since only one 7 foot gauge track had been laid down, and with no turn round facilities at each end, trains would have had to run backwards in the reverse direction. As mentioned earlier, the OWW were required to lay down mixed gauge track on some of the network but the OWW themselves only ever used standard gauge engines and rolling stock. The unwanted broad gauge track was finally removed in 1858.

# Chapter 14

## *Dr. Beeching's railway reshape*

*Dr. Richard Beeching*

It is more than half a century since Dr. Richard Beeching was unleashed on our railways, axing over a third of the existing rail network, closing numerous stations and causing widespread redundancies amongst the workforce. So who the devil was he and why was he allowed to dismantle our rail network in the way he did? In the two books which deal exclusively with this subject, one by Richard Hardy and the other by David Henshaw, each take a diametrically opposite view of the Beeching affair. Yet on one point they were both agreed – that the railways were in bad shape long before Beeching ever appeared on the scene.

Immediately prior to the start of the Second World War, our railways were in the hands of the four railway giants (The London and North Eastern (LNER), the Great Western (GWR), The London Midland and Scottish (LMS) and the Southern (SR)) and none were making a profit. The financial situation worsened as a consequence of the war. During hostilities, all four railways were brought under Government control and became a primary target for the Luftwaffe. Consequently, by the end of the war the railways were in a mess with 2500 miles of track urgently needing repair and more than 8000 locomotives and 12000 passenger coaches needing replacement. This was the situation at the time of railway nationalisation in 1947. The incoming (Labour) Government took the view that what was needed in the first instance was an overarching policy for all forms of transport and created the 'British Transport Commission' (BTC) to oversee its implementation. Its first priority was to salvage a working railway from the remnants of antiquated rolling stock and war-battered infrastructure the government still had at its disposal. This came at an extortionate price at the hands of the 'Big Four' railway companies who, it might have been supposed, would have jumped at the chance of offloading the white elephants in their ownership. With no money to spend, the nation had to wait until the 'You never had it so good' of the Tory government of a decade later before a plan was formulated to modernise our railways. As summarised this was intended to:

*'modernise equipment without making any 'basic changes in the scope of railway services or in the general mode of operation of the railway system.'*

Predictably, the additional costs imposed by the plan only generated more debt for an under-resourced industry. One major component was to replace steam locomotives with cheaper diesel and electric locos; a laudable aim yet only implemented in a half-hearted manner. The problem was that coal was much cheaper than diesel, which needed to be imported, so the interim solution was to continue building steam locomotives, which utilised the abundant locally resourced fuel. In consequence, steam locomotives continued to be built right through the fifties.

Enter Dr. Richard Beeching; in appearance a tubby 'Captain Mainwaring' look-alike, with no previous railway experience, he was head-hunted from I.C.I by Macmillan's government to apply the harsh economics of industry to this essential public service. His previous employers included the Ministry of Defence, for whom he once designed munitions. This was useful experience as he was about to blow the rail network apart. His recruitment was preceded inauspiciously by an innocuous 1960 statement on rail issues presented to the House by Harold Macmillan. A smoke and mirrors affair, on the face of it Macmillan's speech merely addressed an ongoing railway pay review. However, it ominously, contained the following paragraph:

*'the industry must be of a size and pattern suited to modern conditions and prospects. In particular the railway system must be remodelled to meet current needs and the modernisation plan must be adapted to this new shape.'*

The guiding hand behind the 'shape' was Transport Minister, Ernest Marples. Marples had already amassed a personal fortune building roads – in particular the new M1 motorway – and was no lover of railways. On appointment he deflected complaints about his partiality by divesting himself of the shares he owned in his road building company 'Marples Ridgeway'. This he did by donating his controlling interest to his wife. No conflict of interest there then. It should also be added that the Conservative Party at this time was in hock to the road hauliers, who had contributed massively to Tory coffers prior to the previous election.

Marple's first act as minister was to create a Special Advisory Group (SAG) which, deliberately, excluded any existing or former railway people. The Group was set up under the nominal leadership of Sir Ivan Stedeford but, from the outset, the strings were pulled by Marple's appointee Dr. Richard Beeching. Until recently the workings of the SAG were hidden away in the vaults of the National Archive, marked 'Confidential'. They are now available for scrutiny and make for interesting reading.

The first note in the SAG dossier is a letter from Marples to Beeching *(not Stedeford it should be noted)* setting out in one open ended paragraph SAG's terms of reference which were:

*'to examine the structure, finance and working of the organisations at present controlled by the Commission and then advise the Minister of Transport and*

*British Transport Commission as a matter of urgency how effect can best be given to the Government's intentions as indicated in the Prime Minister's statement to the House of the 10th March'*

Beeching was left to interpret this in whatever way he saw fit and duly advanced a four page document that expanded the group's remit to advising the Minister on;

1. the size and pattern of the railway system appropriate to the present and foreseeable needs of the country;
2. the general soundness of the Commission's modernisation proposals;
3. the changes necessary to the Commission's financial structure;
4. the changes in the commission's organisation.

Beeching made clear on page 3:

*'reference is made to the railways only because they represent the major part of the problem'*

Thus SAG were obviously not there to take a holistic approach to the future of transport, only to reform the railways already identified as *'the major part of the problem'*. The Group reported back between July and September of 1960 with eight recommendations:

1. 'pause' the railway Modernisation Programme and stop any associated new projects,
2. set up a 'transport review body' to assess how the development of other types of transport impacted on railways,
3. allow only those railway projects nearing completion to continue,
4. overhaul the BTC finances,
5. immediately require the BTC to,
    a) calculate what could be gained by raising fares
    b) review all existing capital expenditure,
    c) produce a time dated programme for rail closures,
    d) appoint economists to their Board,
    e) require each region to submit regular account/balance sheets
6. remove the requirement of the railways to manage in-house (e.g. 'parcel' and 'bus') companies and 'allow' them to dispose of un-needed assets such as surplus land,

7. stop any further electrification programmes of the rail network,
8. disband the BTC and replace with an alternative structure that included a specific 'British Railways Board (BRB).

The resulting white paper, titled 'Proposals for Reorganisation of the Nationalised Transport Industry', appeared in November the same year. This concentrated on Recommendation '8' and proposed replacing the Transport Commission with a 'British Transport Council' (BTC), which would report directly to Marples but have no financial or executive powers. Railways would be separated from other forms of transport and have their own Board. The proposals were laid before Parliament in June 1961 and the subsequent Transport Act received royal assent the following year. Before this took place, however, Marples appointed Beeching to head the old BTC and clear the way for change. Beeching duly oversaw the complete restructuring of the railway network. It took him less than a year to produce his definitive 'Plan', by which time he had been appointed chairman of a newly created British Railways Board (BRB). The subsequent Beeching Report would seal the fate of our railways for ever. It has been argued over ever since its appearance in 1962, with its recommendations contested by trade unions and the general public alike; the media predictably divided along party political lines. All nevertheless agreed that the recommendations, if adopted, would irrevocably change the face of our rail network. To the astonishment of everyone, with the possible exception of the roads lobby, Beeching advised the closure of a third of the country's railway stations plus around 5000 miles of track. His rationale was to concentrate passenger and freight services on those lines most used and therefore currently generating most revenue.

Surprisingly, in retrospect, opposition to his 'Plan' was muted at best at the time and despite some half-hearted protests from railway trade unions and the odd line-specific challenge by prominent and knowledgeable campaigners, the 'Plan' would be more or less implemented as proposed.

So, assuming the overall purpose wasn't just a roads lobby hatchet job, did Beeching's railway 'reshaping' have any positive benefits? Well, his remit had been to make the railways pay so, on that criterion alone, he failed miserably. Much play was made of the reduction in the railway's annual deficit, which did indeed reduce the year following the Plan's implementation (mainly through the sale of railway assets such as hotels, shipping and land) yet even Beeching's advocate, Hardy, conceded that in subsequent years:

*'the target of putting the railways on to a profitable basis remained as formidable and elusive as ever.'*

Losses increased year on year with the residual skeletal network increasingly falling back on Government subsidies. Some have argued over the subsequent years that Beeching acted in the greater interest of the nation's railways, cutting away the fat in order that our national railway would emerge leaner and fitter[7]. Nevertheless, as Henshaw and many others were to point out, most of the unprofitable routes he axed would have paid their way even at the time, if simple cost reduction methods had been applied (e.g. by reducing overmanning, converting under-used stations to unstaffed halts, introducing automatic level crossings, using cheaper motive power such as railcars etc. etc.). In addition, many of the minor lines closed were originally designed to connect developing urban centres (such as Walsall and Wolverhampton). At worst they therefore had a significant role to play, even if only as feeders to main lines or metro networks. From this standpoint, the changes were short-sighted at best and more than thirty of these axed lines have subsequently been (expensively) reinstated just to accommodate population growth.

It could also be argued that some of the lines Beeching closed would have provided much needed alternative links between major cities. One obvious example is the picturesque Waverley route between Edinburgh and Carlisle, of which the few miles from Edinburgh to Tweedbank have recently been reinstated. The debate continues today over the original decision, which had the effect of isolating some large population centres in the southern Scottish lowlands, such as Kelso. A similar argument could also be made for the retention of the east-west route that once connected Oxford and Cambridge but, just as importantly, avoided rail travellers having to go through London.

A major loss was the trunk route, formerly the Great Central line, from London to Sheffield via Nottingham. This, with its terminus at

---

[7] Particularly Hardy in his biography 'Beeching Champion of the Railway?'

Marylebone, was the last main line to the north built in this country. Although deliberately (according to Henshaw) underused prior to the Beeching cuts, it would have been a vital asset today. Given the furore over the high speed link between London and Birmingham, the Great Central, if it still existed, could have deputised as an alternative route north while the existing main lines were upgraded. This would have rendered the construction of a completely new line unnecessary, saving the nation's purse to the detriment of fat-cat investors – or am I just being cynical?

It was never part of Beeching's remit to consider the social consequences of line closures although it is doubtful, given his personality, that it would have made any difference. A few lines identified for closure, but considered socially important if unprofitable, were subsequently reprieved by the incoming Labour government (e.g. the former Settle and Carlisle). Indeed, Labour based part of their election campaign on undertaking a sympathetic (yet ultimately hollow) review of the rail closure programme. The Government was well aware that users of rural branch lines, especially in Scotland and Wales, would be most affected by the Beeching axe, with the consequence that some less-populated parts of the country are now almost inaccessible by rail. Labour, despite making all the right noises prior to election, did little to reverse or slow the closures once they gained power. Unfortunately, the 'Sir Humphreys' of the Civil Service, with whom the incoming ministers were forced to deal, were broadly in favour of the closures. Their objectivity in providing advice was therefore extremely questionable; the Transport Ministers Permanent Secretary, Thomas Padmore, for example later took up a lucrative directorship with the RAC.

Not all was lost. Some minor sections of the abandoned network were salvaged by rail enthusiasts. Following a pattern established by the Bluebell Railway in Sussex, fragments of the branch line network, such as sections of the Llangollen to Barmouth line, became privately operated 'heritage railways', although it must be said with varying degrees of profitability.

With respect to freight, Beeching noted that a third of the rail network carried just 1% of the total tonnage and advised abandoning all these unprofitable lines. The inevitable effect, to the delight of the men who bank-rolled the Tory party, was to transfer goods haulage to the roads, with the consequent congestion and pollution issues that resulted. Beeching, in fairness, had also suggested the creation of a freightliner network to link all the major coastal ports to London; a move vigorously opposed by the road lobby. Even this small measure, however, couldn't compensate for the irretrievable loss of short haul and long distance freight haulage across the UK. One must therefore assume that the relentless expansion of our road system: the creation and maintenance of new motorways etc. and the ongoing crippling costs to the taxpayer, would have been unnecessary if the existing rail system had not been so myopically devastated.

*Llangollen station, now the terminus of the Llangollen Railway but formerly part of the national rail network.*

As mentioned, the biggest losers of the 'reshape' were isolated rural communities. From now on they were expected to rely on infrequent and unreliable bus services, subsidised by the tax payer, which often ceased to exist after subsidies were withdrawn. The net effect was to render such communities no public transport at all. The removal of railways that served agricultural towns and villages also meant that articulated lorries would jam traffic and pollute the air in tourist hot spots such as Barnard Castle. A few rural lines survived. These were so emasculated by Beeching that they are now of limited value to the communities they serve. One such example is the historic coastal port of Whitby. To travel south from Whitby to London by train involves a fifty mile detour north to Middlesbrough. Many of the discarded rural lines could have provided alternative (and often more scenic) routes for travellers between cities. Their loss has contributed to the desperate overuse and overcrowding of the lines that survived, not to mention the delays caused by track maintenance since there are no suitable alternatives for trains to fall back on.

On top of the all this, there are also safety considerations. Although road fatalities, even in Beeching's day, were much greater in numbers than rail deaths, the social impact of expanding road haulage, as opposed to safer rail travel, was not considered in the reshape plan. Railways had, and still have, the best safety record of all forms of transport so we must assume the annual toll of road casualties might have been substantially reduced if our original railway network had been retained. The adverse road/rail safety comparison was known at the time but never taken into consideration in the Beeching report.

This combination of public safety, lower pollution and lower road congestion has resulted in the creation of 'metro' networks linking city centres to the suburbs. The introduction of the urban metro, often, ironically, utilises the same lines that Beeching axed (e.g. Bury to Manchester). Bus stop type halts now substitute for expensively manned stations along the old routes. This simple cost cutting solution to the staffing of rural branch lines was well understood in Beeching's day, having been pioneered by 'tramways' and 'light railways', such as those operated by Holman Stephens described elsewhere in this book. That this well-tried and obvious model was not considered is a glaring and notably suspicious omission.

Following the Beeching Report, 100,000 railway workers lost their jobs yet the railway's annual deficit increased from £65 million to £73 million, between 1963 and 1965 (although wages for those retained improved – including, it should be noted, those of Beeching's main apologist Richard Hardy).

*Corfe Castle station, now on the Swanage Railway but formerly part of the national rail network*

So how do we summarise the Beeching effect on our railways?

Well, as noted, on purely financial terms it was a failure. It never achieved the objective in the Macmillan statement to the House of the 10th March 1960 that the railway would be *'of a size and pattern suited to modern conditions and prospects'*. By destroying a coherent, if unprofitable, railway system Beeching handed the nation's future transport arrangements to the road lobby with the predictable consequence of pollution, congestion and road fatalities that we see today. By destroying the rail feeder network, the future development of the railways as a viable competitor to roads became impossible. Enfeebled by a lack of investment by successive governments, the railway became ripe for privatisation – an issue assiduously pursued by John Major's fag-end Tory government – culminating in a final act of vandalism, fragmentation and exploitation an incoming Labour government of 1997 could have stopped but, like their predecessors in the sixties, did nothing to reverse. We live with the consequences. Our railways are more expensive, more crowded and as state subsidised as our European neighbours with the difference that any profits over here are used to pay dividends to shareholders.

It is sad that an affordable railway, designed to serve the nation, has not been seen as a valid and necessary public service, like the NHS, and owned by the country's taxpayers. As our armed forces get tied up in ever increasing numbers of wars, to maintain the flow of oil that keeps lorry wheels turning, the destruction of a more energy-efficient alternative, bequeathed to us by our forebears, seems criminal at best. This is the true legacy of Beeching's 'reshape' of our railways.

**Notes to Chapter 14**

Lest I be considered biased in my appraisal of the good doctor I quote from Wolmer's book, 'Fire and Steam':

*'Overall the closure was a haphazard process based on flawed logic and the assumptions by Beeching about the lack of profitability of many lines were proved wrong.'*

# Chapter 15

## *Nigel Gresley's Soccer Specials*

*Gresley A4 class 4-6-2 'Sir Nigel Gresley' at the National Rail Museum at Shildon*

I look forward to the reappearance of 'Manchester United', the locomotive I might add, not the football team. I always had a soft spot for the B17s named after football teams. This has something to do with the fact that I never saw any of the engines in the dim distant train spotting days of my youth. To be truthful, I never saw any Southern or Western engines either but somehow that never bothered me much. No, the truth is there was something special about the B17s and this was purely because of their football connection. Being equally mad about soccer and railways, I always wanted to 'cop' a steam locomotive called 'Newcastle United', 'Middlesbrough' or even, at a pinch, 'Sunderland'. Being a Hartlepool United supporter, I drew the line at 'Darlington'. But Darlington F.C. bias aside, I don't think there was anything unusual in this. I suspect that even the technological crazy youth of today would think there was more street cred in seeing a loco called 'Tottenham Hotspur', say, than 'Cuddie Headrigg 'or 'Luckie Mucklebackit' – no disrespect. When I also discovered that two of the B17s were also streamlined, like Gresley's 'streaks' (as seen above), the need to meet one became desperate. Sadly, this was an unrequited love. There were two problems: 1) none of these engines worked in County Durham where I lived and 2) all the horses in this particular stable went to the knacker's yard long before I was able to raise sufficient disposable income to go out and find them.

For the uninitiated, a few words about the B17s and, since I have never been able to summon much interest in technical details, I apologise to those who want to know, for example, the specific dimensions of a B17 boiler. Such enthusiasts could do worse than read Peter Swinger's book (see bibliography) which gives a sequential summary of each locomotive in the class. For the less technical, all you really need to know is that these handsome 4-6-0s MTs engines were designed by Nigel Gresley and built between 1928 and 1935 in Glasgow, Darlington, and Doncaster and used primarily on the former Great Eastern lines south of the Wash. As such, unfortunately, they rarely ventured north of Sheffield. Additionally, and also unfortunately, only two of the class were ever streamlined, although this did provide those particular engines with a certain cachet. The lucky pair were modified for fast working on the 'The East Anglian' express between Norwich and London, it being anticipated that they would be a showcase for high speed travel between London and the eastern counties. In the event, both the image and the hype didn't match up to the reality and the streamlining turned out to be a vanity exercise. The problem was that speed restrictions on an already overcrowded main line made continuous fast running virtually impossible.

The 4-6-0 'streaks' were initially painted apple green, to distinguish them from the blue and grey Gresley A4 Pacifics they so physically resembled. One was named 'City of London', in the process losing its allocated football name[8]. Its sister was called 'East Anglian' for reasons which will shortly become clear.

*Streamlined Gresley B2 class 4-6-0 'East Anglian', as new*

No further attempt was made to streamline the rest of the B17s so this experiment must have been deemed a failure. Nevertheless, the streamlining survived the Second World War, during which time the engines were painted black. Following nationalisation, they were repainted in the more subdued but more familiar British Railways Brunswick Green. By now the streamline cladding was proving more trouble than it was worth in terms of engine maintenance, and was finally removed in 1951 along with the haunting 'chime' whistles. I am indebted to James Woodrow for a potted history of one of the 'streaks'. LNER locomotive number 2859 emerged new from Darlington Works on the 11th June 1936 and was allocated the name 'Norwich City', a name associated with the engine for just the first few months of its existence. It was renamed 'East Anglian' to coincide with the inception of the eponymous express train but a name change of some sort was under consideration from the outset since 'Norwich City' was already too emotive for a loco that regularly worked the territory of local rivals Ipswich Town, recently promoted to the Football

---

8   Both engines in their pre-streamlined form bore the names of football teams. LNER 2870 was both 'Manchester City' and 'Tottenham Hotspur' at some time. LNER 2871 also started life as 'Manchester City'. Confusing isn't it?

League. It has been suggested that the streamlined version of 2859 carried a 'Norwich City' nameplate for a few weeks but this seems unlikely given the underlying reasons behind the streamlining. It is therefore of no surprise that, as far as I can tell, there are no publicity photographs of a 'Norwich City' streak, whereas there are several publicity shots of the streamlined 'East Anglian' looking polished and pristine and obviously fresh from the works. The probability is that the football name, if it ever existed, was removed before the fancy cladding was added.

*Naming ceremony for Gresley B2 4-6-0 'Sunderland'*

The name 'Norwich City', nevertheless, was not lost. It was transferred to another B17, number 2839, where it remained until the engine was withdrawn from service in May 1959. The curved nameplate above the players' tunnel at the Norwich City football stadium today is the one taken from 2839 (B.R. no. 61639)[9]. It was officially unveiled to the club's supporters in October 1959, following a friendly against FA Cup holders Nottingham Forest. Originally located in the Main

---

9  It's possible that the curved nameplate originally adorned the pre-streamlined LNER No.2859

Stand, it moved to its present position after the stand caught fire in 1984.

Notwithstanding the way the names were swapped around, change was to be a defining feature of the B17s. During the course of their lives, Gresley's original design was regularly messed with. Major rebuilds of the engines were undertaken; the boilers of all, for example were modified as late as 1958, just two years before the seventy three 'Sandringhams'[10], as they were now known, were scrapped, a sad loss to both steam heritage and football.

Although the claim to fame of the B17s, at least from my personal perspective, was their tangential football connection, in fact only twenty five were given the names of Football League clubs. All of these, unsurprisingly, were built at Darlington in the football-mad North East, with the intention of using these engines to front soccer 'specials'. In retrospect, the powers-that-be can't have thought this idea through. Restricting the use of a locomotive with a name such as 'Leicester City', for example, to soccer specials involving only that particular football club would have resulted in serious underuse, yet the alternative, involving the same engine perhaps fronting a train full of tanked up supporters from local rivals Nottingham Forest, would have cause riots. In many instances the engines weren't even allocated sheds close to where their eponymous football club was located so the logic behind the allocation of football names was fundamentally flawed. Nevertheless, the principle of naming locomotives after football teams was a breath of fresh air in an industry that mostly genuflected to the ruling classes and, if only from this perspective, these engines deserve special praise. In no particular order the football B17s were[11]:

**Tottenham Hotspur, Norwich City, Arsenal, Sheffield United, Grimsby Town, Derby County, Darlington, Huddersfield Town, Sunderland, Middlesbrough, Leeds United, Doncaster Rovers, Hull City, Sheffield Wednesday, Manchester United, Everton, Liverpool, Leicester City, Nottingham Forest, Bradford, Bradford City, Barnsley, West Ham United, Newcastle United and Manchester City.**

---

10   After the first locomotive in the class LNER number 2800 'Sandringham'.
11   Randomly arranged to avoid upsetting rival football fans.

Nearly all had curved nameplates located over the central driving wheels; the team name underscored by a brass football painted in the colours of the club. The exceptions to the rule were the two streamlined engines which had straight nameplates located over the leading wheel bogie like the A4s they so closely resembled. The nameplates of all the B17s and their companion B2s, however, were semi-permanent features. Both Newcastle United and Manchester City, for example, were renamed within months of entering service. Manchester City was the unluckier of the two; the 'Manchester City' nameplate was initially transferred from engine 2870 to engine 2871 and then discarded altogether in favour of the name 'Royal Sovereign', so named because she headed a royal train to Sandringham. 'City' at one time also had the dubious distinction of seeing its name changed from 'Manchester City' to 'Tottenham Hotspur', which can't have impressed either set of football fans.

Having a name that reflected the day-to-day experience of 'working class' people was rare for steam locomotives. It hadn't occurred to me how rare until I came to write this piece. Acknowledgement of the life and culture of the majority of the country's population hardly occurred at all over two hundred years of public railways – even most of the B17s/ B2s were named after stately homes.

To gauge how rare such nomenclature was, I trawled through my old Ian Allen ABC books and assigned social category labels to each listed name. I may have missed a few, or made the occasional counting error, as my eyelids were drooping long before the end. The plan was to identify engine names with direct relevance to the lives of manual workers, after all, it was working class people that built them and who comprised the bulk of the railway workforce. How many pre-nationalisation locomotives, amongst the numerous Shires, Granges and Halls, I wondered, were named after people or places familiar to the working man? I therefore scanned the lists looking for collieries, dockyards, shipyards, steel works, chemical works, any other engineering works, cotton mills, farming or other agricultural or food based industries, state schools, hospitals[12], churches, public buildings (including museums and libraries), theatres, authors, musicians, music halls, famous comedians, newspapers, comic book characters, singers, film and radio stars, sportsmen and

---

12  I don't count the public school 'Christ's Hospital'

sportswomen, sports in general (other than field pursuits), labour politicians, working men's clubs, trade unions (including railway trade unions) philanthropists, charitable institutions, scientists, philosophers, or engineers otherwise unconnected to railways. The answer essentially was none.

Of the steam locomotives built by the pre-nationalisation 'Big Four', more than 45% were named either for past or present members of the aristocracy or, failing that, their homes and/or leisure pursuits (particularly fox hunting[13]). London Midland and Scottish (LMS) were the least deferential with 13% of their engines named after Princesses and Duchesses (although these tended to be figurehead engines in the fleet) whilst in terms of forelock tugging no-one tugged harder or more often than the Great Western Railway with a staggering 92% of all their engines with names designed to appeal primarily to landed gentry. The charitable view is that these names were chosen because they created a classy image for GWR rolling stock, if only by association, as nearly all were named after Kings, Granges, Castles, Manors, Courts and Halls. The less charitable view is that by so-naming their locomotives the executives of the GWR were hoping that doors would open for them in the stately homes after which the engines were named.

Of course, amongst the Kings, Queens and minor Royals that dominated the pre-war companies' loco fleets, there are inevitably a smattering of other names; there are only so many royals to go round after all and, when it came to providing other names for engines, each Company displayed its own specialist area of interest. For the LMS it was the military (The Great War, fighting ships, regiments etc.) whilst the Southern Railway preferred local place names and Celtic mythology. The LNER had a penchant for animals (particularly racehorses): and the GWR – well, apart from the occasional saint or place name, not much else to be honest. As noted previously, two thirds of the B17/2s were also named after country residences belonging to the well-heeled, some of whom – such as the 19th century occupiers of 'Wynyard Park' – not only hated railways but did everything in their power to prevent them crossing their land.

---

13   There are three times as many locomotives named after 'Hunts' as Association Football Clubs

Of late, locomotive names have become more egalitarian; it is hard to imagine a pre-nationalisation steam locomotive, other than perhaps the odd works engine, being named 'Hartlepool Pipe Mills', for example. But whether or not you prefer this type of name to, say, 'King George III', depends on your background and/or historical interest in mad monarchs, but at least it is now possible to see some variety.

Getting back to football, what did the existence of a locomotive named after their football club mean, if anything, to the clubs concerned? Well, a measure of this can be ascertained from determining the fate of the relevant nameplates once the engines were scrapped, since each of the League clubs involved were offered either one or both plates. Surprisingly, not all football clubs took up the option, even though it cost them nothing to do so. Prominent refuseniks included such big names as Manchester United and, of those that did, the majority didn't hang on to the original plates (I should add a caveat here in that the response to my request for information from the clubs themselves was patchy at best). I found, to my frustration, that the bigger the club the more difficult it was to establish contact. The primary aim of websites of most of the Premier and Championship clubs, it seems, is for making commercial transactions. There is no problem, for example, applying for match tickets or obtaining squatting rights to hospitality suites or the acquisition of a formidable list of club merchandise, but there is little provision for losers like myself wishing to seek answers to non-financial queries.

I should point out, in fairness, I hadn't expected a response from clubs no longer in the Football League such as Darlington, Bradford Park Avenue and Grimsby Town and was indeed surprised when the contact for Grimsby Town turned out not only to be a railway buff but knowledgeable about the B17s. This was a bonus since, on the face of it, the choice of 'Grimsby Town' for a locomotive name wasn't obvious; Grimsby were hardly a leading footballing light even in the 1930s. Additionally, the rest of the names selected by the LNER represented just a fraction of the teams in the then Association Football League. The reason for the particular names selected was that the LNER planned to name all these engines after the top football clubs of the day, which at the time included Bradford P.A.. These clubs were also the best supported in the country and football specials involving them made obvious financial sense. A last minute decision was also made to name four of the engines after the 1936 FA Cup semi-finalists, which by chance that year included Grimsby Town, hence its appearance in the list. The naming of the similarly anomalous 'Doncaster Rovers' and 'Darlington' is easier to understand, given the location of the works where these engines were built.

*Gresley B2 class 4-6-0 Manchester United*

Of the clubs that replied, to my questionnaire only Tottenham Hotspur had retained both the B17 nameplates. Less financially secure clubs, such as Doncaster Rovers, were obliged to sell theirs and purchase replicas which are now displayed in clubhouse bars and reception areas, although in a couple of instances the nameplate or replica is positioned prominently over the players' entrance tunnel, Norwich City being one such example. Nearly all the clubs have at least a picture of their own locomotive on the walls of the hospitality lounges and the majority of my correspondents expressed regret that more interest hadn't been taken, at the time the engines were withdrawn from service. Given today's market value of steam locomotive nameplates this is unsurprising.

Yet from the perspective of a B17 enthusiast, all is not lost. Whilst the nameplate of the original 'Manchester United' adorns the walls of the National Railway Museum at York, there is hope that the name will also one day reappear on a working steam loco. Two 'new-build' projects have been established with the aim of constructing a B17 from scratch in the manner of the Class A1 'Tornado'. One of the projects involves the construction of a facsimile 'Manchester United' and the other, like Tornado, involves an addition to the class provisionally named 'Spirit of Sandringham'; the name harking back to the first engine in the original series. Like Tornado, 'Spirit' will be allocated the next unused sequential BR number '61673', following on from the last engine in the original BR series, 61672 'West Ham United'. Due consideration was apparently given to naming 'Spirit' after a current football club but the idea was rejected as being potentially too divisive; the football supporter of today being even more tribal than its 1930s counterpart. No component of the original B17s survived the scrap man's torch so 'Spirit of Sandringham' will combine the tender of former Gresley J39 '64961' and the chassis of Gresley B12 '61510' along with anything else cobbled together, dating from the time the Sandringhams were built. The specification, of course, is to be updated in line with present standards of health and safety etc. but the original engineering drawings will be used as much as possible so that 'Spirit' most closely resembles the later designs Gresley used for his 'football' locos.

The 'Spirit' project is still at an early stage, as is the new 'Manchester United', the latter being assembled on the site of the Mizens miniature railway at Woking where the engineering team have acquired a Gresley tender and fabricated a cab. Both projects require significant public support if they are ever to achieve completion. It seems to me it would be a nice gesture if this Premier League club, who seem happy to shell out millions for footballers, would also offer to help, if only for the positive publicity the Club would obtain from so doing.

I wish the new-builds well. With their help I may yet achieve my boyhood desire of seeing a B17 in steam. In my imagination I still dream that one day I will be standing on the platform of the North York Moors Railway station at Grosmont, and hear the sound of a 'chime' from the adjacent tunnel. Bursting forth, in a cloud of steam, will be a new-build B17. It will be streamlined, painted Brunswick Green and sporting a blue and white nameplate bearing the legend 'Hartlepool United'.

I live in hope.

*Players take the field at Norwich City beneath the ex B2 nameplate over the players' tunnel*

**Notes to Chapter 15**

As of the time of writing these notes (August 2016) little progress seems to have been made on the new-builds. As is always the case, money is the big issue. Following the construction of 'Tornado', and the popularity of the recently rebuilt 'Flying Scotsman', this seems, at first sight, difficult to comprehend, until you consider how many such projects are in various stages of completion throughout the UK. What resources are available for such projects are spread wafer thin.

# Chapter 16

## *William Hedley's Wylam Dillys*

*Portrait of William Hedley as a young man*

On his way to school in Wylam, from the nearby village of Newburn, William Hedley would have passed a small miners' cottage, right beside the waggonway between Wylam Colliery and coal drops on the River Tyne at Lemington. Outside the cottage there would have been children playing, perhaps even dodging the wheels of the horse drawn coal wagons or chaldrons that trundled past. One of these 'bairns', either playing there or perhaps opening and closing the nearby crossing gate, could well have been George Stephenson whose family occupied one downstairs room in the cottage. Their paths would cross many times over the coming years.

William Hedley was born at Newburn on 13th June 1779. His father, also William, was a grocer from Throckrington in Northumberland; sufficiently well off to afford the education of his son at a small private 'school' in Wylam village.

It is three miles from Newburn to Wylam and it must have been a long walk for young Hedley who was severely asthmatic. Nevertheless, under the tutelage of his teacher, Mr.Watkin, Hedley blossomed. Hedley was particularly good at mathematics which led to his apprenticeship as assistant 'Viewer' at Walbottle Colliery; the term 'Viewer' being the name applied to colliery managers, whose role could vary from foreman to that of Resident Engineer. The Resident Engineer at Walbottle, in fact, was the future locomotive manufacturer, and rival to Robert Stephenson and Co, Robert Hawthorn and a friendship between Hedley and Hawthorn developed that would stand them in good stead in later years. It is a testament to the ability of the youngster that, by the age of 21, Hedley had become the Viewer at Walbottle.

Following his marriage to Frances Dodds, Hedley began investing in mining and ship building, which perhaps suggests financial support from his wife. He bought shares in a lead mine at Blaghill, near Alston, and then became the mine's acting general manager, being sufficiently well thought of that when he eventually left they gave him a tea and coffee service made entirely from silver extracted from the mine. In 1805 the family moved to Wylam and leased a house belonging to local mine owner Christopher Blackett. Blackett is an interesting character who deserves his own place in the railway hall of fame. He had inherited a failing colliery business on the death of his spendthrift half-brother Thomas and, despite having no previous mining experience, re-built the family business and paid off the outstanding debts. Shortly after acquiring Wylam Colliery, Blackett contacted Richard Trevithick, the engineer and steam locomotive pioneer, and asked the Cornishman to build him a locomotive for the colliery, probably the first such commission in history. Britain was at war with France then and horses and horse fodder was in short supply so Blackett's motive was in finding a cheaper means of moving his coal to the docks on Tyneside. Trevithick, unfortunately, was otherwise occupied and declined the offer. However, he did agree to oversee the building of an experimental engine at Whinfield's Foundry at Gateshead. Named the 'Newcastle', it unfortunately never left the factory because Blackett thought it too heavy for the Wylam tramway's wooden rails. Nevertheless Blackett could still see the potential of steam and set about improving his waggonways to make locomotive use more practical.

William Hedley moved to Wylam because Blackett, fresh from his locomotive experiment, offered him the Viewer's job at the Colliery and Hedley's first task was to replace the existing wooden waggonway with cast iron 'plate' rails[14]. This took him more than three years and only when that was completed did his employer turn his attention once again to steam traction. Hedley was then instructed to build a steam locomotive for the colliery and to assist him he was provided with the services of the colliery engine-wright Jonathon Forster and the young works blacksmith, and future locomotive pioneer, Timothy Hackworth.

Looking back two hundred years, it is hard to imagine the controversy that then raged over whether it was possible for a heavy vehicle to get adequate traction on smooth metal rails. Empirically, we might understand why. One of the reasons smooth metal rails were laid down in the first place was to minimise friction between wheel and rail and thus overcome resistance to moving heavy loads. In the days when haulage on the tramway amounted to just a horse with a couple of loaded trucks, lack of friction between wheel and rails wasn't seen as a problem, but as soon as the transport medium became a multi-ton engine there seemed a conflict of interest between the frictionless nature of the rails/wheels and the force required to make the engine move forward. In consequence, the common belief

---

14  Plate rails were angled iron pates which were laid on top of wooden rails to minimise damage to the wood. Unlike today's metal rails they allowed the use of wagons with conventional road, as opposed to flanged type wheels.

was that locomotive wheels would not grip under load. The strange thing is that Trevithick's experiments, a decade earlier, had already demonstrated that, except under very wet conditions, this wasn't an issue The actual problem was that Trevithick didn't grasp the concept himself. As a result of this, Trevithick went out of his way in his patent applications to suggest unnecessary additions, such as rack and pinion devices, to ensure adhesion between wheel and rail. As we now know, other than in the very worst weather conditions, this was a waste of effort since friction is actually a function of both the relative smoothness of the contacted surfaces *and* the pressure applied between them. Therefore, the very weight of a locomotive is normally sufficient to prevent wheel slip.

*The Hedley test carriage*

Nevertheless, in 1811, Christopher Blackett was unconvinced and, rather than commit to building a smooth wheeled locomotive that might not work on his waggonway, he asked Hedley to undertake experiments to evaluate the principle. Hedley, with assistance from Hackworth and Jonathon Forster, duly constructed an ingenious device for carrying out wheel adhesion trials, called the 'test carriage'. It consisted of a four wheeled coal waggon, which had been modified to accommodate hand cranks attached to a fly wheel, linked by cogs to the wheels. Four men were stationed on platforms on each side of the 'carriage' and by turning the crank handles the machine was able to move backwards and forwards.

The experiments were carried out in secrecy. Blackett knew that if the experiments were successful he could use locomotives on the tramway and significantly lower his transport costs, thereby increasing the profit margins on his coal. He wasn't therefore prepared to give anything away that could assist his rivals. It is a measure of how secretive the work became that the trials were conducted, not at the colliery itself, but on a piece of specially constructed track in the grounds of Blackett's own home, Wylam Hall. In keeping with this cloak and dagger approach, work only started when it got dark. The exact form the experiments took is best explained using Hedley's own words. In a letter to a Dr. Lardner, published in the Newcastle Courant newspaper of 10$^{th}$ December 1836, he said:

*'The carriage was placed upon the railroad and loaded with different parcels of iron, the weight of which had previously been ascertained: 2, 4, 6, etc. Loaded coal wagons were attached to it, the carriage itself was moved by the application of men at the four handles and, in order that the men might not touch the ground, a stage was suspended from the carriage at each handle for them to stand upon. I ascertained the proportion between the weight of the experimental carriage and the coal wagons at that point when the wheels of the carriage would surge or turn round without advancing it. The weight of the carriage was varied and the number of wagons also with the same relative result.'*

In other words, Hedley loaded the test carriage and measured the weight of a loco capable of pulling specified numbers of loaded coal wagons without generating wheel slip. How many experiments were conducted we don't know, but Hedley describes the experiments as being 'on a large scale', the outcome being that:

*'...the friction of the wheels of an engine carriage upon the rails was sufficient to enable it to draw a train of loaded coal wagons.'*

One assumes, for the sake of scientific reproducibility, he always used the same men in the same location on the carriage. The results of the experiments were never published nor is there any mention of them in the subsequent locomotive patent Hedley applied for. In fact, in the patent application Hedley resurrected the already redundant idea that his proposed locomotive could be fitted with cogged wheels to minimise wheel spin. However this may have been a red herring designed to mislead Blackett's competitors. Regardless of this, having demonstrated the adhesion principle, at least to his employer's satisfaction, it was left to Hedley to construct the Colliery's first steam locomotive.

With no previous expertise to draw upon, Hedley sought assistance from Thomas Waters, now the Whinfield Foundry manager, who previously worked on the ill-fated 'Newcastle'. The Wylam 'test carriage' provided the basic frame on which the locomotive would be assembled. Waters, who oversaw the construction, made the boiler at the Gateshead foundry. The water in the boiler was heated by a single flue from the firebox and the steam generated fed to a single piston which turned the cogged fly wheels Hedley had previously used in his 'test carriage' experiments. The rest of the locomotive's component parts were manufactured in Hackworth's blacksmith's shop, with Hedley providing Hackworth (and perhaps Forster) with thumbnail sketches of what he was after, which he drew in chalk on the door of the blacksmith shop. The only tools the men had were a small hand lathe, some hammers, chisels and files. Nevertheless, despite its Heath Robinson origins, the first Wylam locomotive took to the rails some time towards the end of the year 1811 or early in 1812.

Like its successors, it acquired the nickname 'Grasshopper', presumably because of its 'leggy' appearance, and was tested over the two miles of waggonway between Haugh Pit in the centre of Wylam village and 'Street House' where George Stephenson once lived. Pulling five chaldrons of coal it stuttered along, stopping at intervals to draw breath. On arriving at Street House, on one occasion, it gave up the ghost altogether due to lack of steam pressure. The engine had been fitted with a crude safety valve which amounted to a metal weight over a pipe that fed directly into the boiler, the boiler itself being little more than a barrel made out of welded metal strips. As the assembled spectators watched with growing trepidation, Waters loaded additional weights on to the crude safety valve and instructed his terrified assistant to stoke up the fire. Steam pressure, as expected, increased dramatically yet, instead of exploding, as the boiler had every right to do, it somehow stayed intact and, with a shudder, the locomotive lurched forward again and stuttered its way back to the colliery in a series of fits and starts.

Grasshopper was consequently hardly the success Blackett was looking for. It had a habit of running out of steam at crucial moments. For the next twelve months, a number of improvements were made to try and solve the pressure loss yet the one simple modification that might have saved the engine (i.e. the ejection of exhaust steam into the chimney, which would have increased the draught through the fire box), was not attempted. In fact, most of the heat generated in the firebox never got to raise the temperature of water in the boiler. Heat wastage was so great that the locomotive's chimney glowed red hot, rendering the operation of the engine a perilous undertaking for the poor fireman. The sound and fury of the clanking Grasshopper, chimney glowing, was likened by one observer as the 'passing of a roaring meteor'.

Nevertheless, on good days the loco could move up to 6 wagons full of coal from Wylam to Lemington staith without mishap. Unfortunately, this occurred only on good days. More often it broke down and had to be hauled back to the colliery using the horses it was meant to replace. One or two notables joined the crowds of onlookers that watched the engine perform. There was Hedley's old friend Robert Hawthorn and, less welcome perhaps, the regular presence of George Stephenson. George was now working for the mining co-operative known as the Grand Allies at the nearby Killingworth Colliery and was given the task of finding out what exactly was happening at Wylam. Stephenson received daily updates on the progress of the Grasshopper from his friend Jonathon Forster, who assisted in the construction of the machine, but he was soon warned off by Hedley and thereafter was obliged to make clandestine visits to acquaint himself with developments. It is possible, however, that Hedley was well aware that Stephenson was checking what was going on and was happy that the 'Allies' were being advised of the problems encountered with the experimental engine. What Stephenson's employers didn't know was that in terms of locomotive development Hedley had already moved on. As a haulage experiment, the 'Grasshopper' had served its purpose but was soon relegated to a stationary position in the colliery where it served out its days as a winding engine; its demise of little regret to the colliers who had been forced to give up their own time to work on it. The pitmen never knew the engine by any formal given name, if indeed it ever had one, to them it was just the 'Dilly', a local slang term for a cart horse[15].

*Puffing Billy at the Science Museum, South Kensington*

Almost all we know about the construction of Grasshopper's successor, 'Puffing Billy', we can infer from an angry letter from Hedley to a certain Dr. Lardner that Hedley wrote in 1836. What got up Hedley's nose was having to sit through one of Lardner's lectures in Newcastle in which George Stephenson had been given sole credit for inventing the steam locomotive. So what did Hedley have to say about Puffing Billy?

---

15 This may not be true. Although from the North East myself, I have never come across the word used in this context. The Geordie slang word for a horse that I knew was the word 'cuddy'

*'Another engine was constructed; the boiler was of malleable iron, the tube containing the fire was enlarged, and in place of passing directly into the chimney it was made to return again through the boiler into the chimney at the same end of the boiler as the fire was... The engine was placed upon four wheels and went well; a short time after it commenced, it regularly drew eight loaded coal wagons after it, at the rate of from four to five miles per hour.'*

Hedley duly took out a patent on the engine in 1813.[16] We are fortunate we do not have to rely on old drawings to see what Puffing Billy looked like. Both the original and a working replica exist today. At the industrial museum at Beamish in County Durham the replica trundles up and down a short length of track coupled to waggons loaded with museum visitors. In appearance it resembles a sewing machine on wheels. The original Puffing Billy, unlike the Beamish clone, had no brakes and relied on gears and the primitive brakes on wagons to enable it to stop. 'Puffing Billy', unlike its predecessor however *was* provided with two cylinders. These were connected, as was the first Wylam engine, to the wheels via a system of cogs. There were other important improvements, the most innovative (as Hedley pointed out in his letter to Lardner) being the radical redesign of the boiler. In earlier locomotives water was converted to steam by directing hot gas from the firebox straight through a flue within the boiler body into the chimney after which it directly discharged to atmosphere. This was wasteful in terms of heat and expensive in terms of fuel. In Puffing Billy the flue from the firebox ran the length of the boiler, as before, then returned back on itself before the hot smoke was allowed to escape from the chimney. Heat exchange efficiency between fire and boiler water was therefore effectively doubled. This adaption made for a strange look to the engine. Both the chimney and firebox were now at the same end of the boiler so the locomotive needed two men to operate it; a fireman loading coal at the chimney end and the driver working the controls at the other end. One other controversial innovation of note is that exhaust steam was then forced through a constricted tube directly into the flow of smoke in the chimney. This, in effect, was a crude version of the blast pipe seen on modern steam locos, which led to much heated debate as to ownership of the blast pipe principle, something which rages on to this day. The blast pipe

16  The patent itself is a strange affair saying very little about the construction of the locomotive and including no drawings

principle is as follows: the discharge of high velocity steam into the smoke flow in the chimney increases the flow of gas in the chimney and thereby draws more air into the firebox to compensate, thus creating a brighter and hotter fire and hence more steam. At a stroke therefore Hedley solved the problem of maintaining steam pressure, but whether he was aware of exactly what he discovered is still open to question since he made no attempt to patent his discovery.[17]

In reality the crude blast pipe was probably a lucky accident. Exhaust steam from Grasshopper had discharged directly onto the track, wetting the rails and causing wheel slip. Hedley had therefore only arranged for steam to discharge through the chimney to eliminate this problem, whilst at the same reducing the high pitched hiss of steam emissions, which had been the subject of numerous complaints during the original locomotive trials[18]. Nevertheless, the effect this modification had on increasing the heat to the boiler did not escape the attention of another who worked on the engine – the colliery's foreman blacksmith. Hackworth later refined and patent the blast pipe and used it effectively on engines he subsequently built for the Stockton and Darlington Railway.

If the blast pipe was a lucky fluke, the U shaped return flue was a brilliant and original invention. This, more than any other modification, turned Puffing Billy into the world's first successful and reliable conventional steam locomotive. It was a success from the outset. According to Hedley's son Oswald, the new engine:

'conveyed sixteen waggons at the rate of upwards of five miles per hour (and) with the carriages empty from six to seven (miles per hour)... doing the work of 16 to 18 horses.'

Replacing horses was not, it should be said, universally popular with the men at the pit. Amongst their number were horse handlers and stable boys, none of whom would be needed if the engine proved successful. The relative failure of Grasshopper had been viewed with amusement by the stable people whereas the impressive achievements

---

17  He never, for example, used this as a novel feature of his patented engine or even referred to it in his letter to Dr.Lardner.
18  The sound of discharging steam was reputedly so loud that it startled horses into bolting on the adjacent highway.

of Puffing Billy were received with downright hostility. From the first days of steam at Wylam, there were problems with rail damage caused by the weight of the engine but they were to be aggravated by actual sabotage of the rails by disgruntled horse handlers, whose livelihoods were now under threat. Hostilities were put on hold every Friday when the men received their pay. Dressed up to the nines, and with their wives and children in tow, they boarded coal trucks pulled by the hated 'Dilly' and were taken to Lemington where keel boats took them on to the bright lights of Newcastle. The same train picked them up at the end of the day and brought them back. It could be argued, therefore, that it was on the Wylam waggonway that the world's first scheduled locomotive hauled passenger service operated. What condition the day trippers were in after a night on the town and two outings in a coal wagon, however, doesn't bear thinking about.

The regular operators of Puffing Billy were John (Jacky) Bell and John Lawson. We know this because their initials are carved on the firebox of the engine which is now a permanent resident of the Science Museum. These men called the locomotive the 'Dilly', so where did the name Puffing Billy originate? One anecdotal theory is that Puffing Billy was the derogatory term used by the pitmen to describe Hedley, an unflattering portrait of him aged 29, shows him to be a tad overweight, to put it mildly[19]. He was also asthmatic and the nickname Puffing Billy for the Viewer seems plausible. It has been suggested that only much later did the name of the locomotive and that of its designer became synonymous.

It wasn't long before 'Puffing Billy' began chewing up the thin plate rails on the waggonway as its predecessor at Gateshead had done. In consequence, Hedley modified it; spreading the weight more evenly by remounting the boiler on eight wheels and, as an eight-wheeler, it continued to operate until the plate rails were replaced with conventional edge rails in the 1820s.

People came from all over the world to pay court to the 'Dilly'. In 1815 Archdukes Lewis and John of Austria took a boat up the Tyne from Newcastle to Lemington just to see the engine at work. Its favourable reputation must have been particularly galling to that other Wylam

---

19  Not least by the operators of the replica engine at Beamish.

son, George Stephenson, whose own engines were now operational but performing indifferently.

*Wylam Dilly at the National Museum of Scotland, Edinburgh*

Flushed by success, Hedley built two more engines to the same template. The first acquired the name, more appropriately perhaps, of 'Wylam Dilly' and the second was apparently known as 'Lady Mary'. Of the latter we know little, other than its supposed image in an old photograph, which indicated it was an engine similar in every respect to its sisters. This has led to claims that it never existed at all[20]. More

---

20  Most recently by John Crompton of the National Museum of Scotland in a paper presented to the Newcomen Society in 2003. His main argument was that 'Lady Mary' looked too like Wylam Dilly to be a different engine. It does seem likely that the picture is of Wylam Dilly. Lady Mary probably never survived into the days of photography.

likely it had a short life as a working locomotive. One of Hedley's engines was known to have been butchered to provide motive power for a modified keel boat, during a keel-men's strike in 1822. Since Puffing Billy and Wylam Dilly were in daily use, it is possible that this was the fate of 'Lady Mary'. Wylam Dilly and Puffing Billy both survived virtually unaltered into old age, although their original four wheels were restored when eight wheel support for the boiler became unnecessary. The Wylam locomotives would go on to haul many thousands of tons of coal over the years, from both Wylam and, in the case of Wylam Dilly, also at Hedley's own colliery at Craghead.

Despite his achievements, Hedley remains a hard person to like. On paper he comes across as a bumptious abrasive character. It is no surprise therefore that there are no fond anecdotes of 'Gaffer Billy' in the copious annals of the Mining Institute. The further development of locomotives at Wylam fatally stalled when Hedley fell out with his blacksmith. Hackworth was a devout Methodist and refused Hedley's summons to work on the 'Dilly' on the Sabbath. Hedley wouldn't compromise and Hackworth walked out, even though he was happy at the Colliery, was a native of the village and had, on the face of it, no other employment to move to. It is possible that Hackworth's brother Thomas, also a blacksmith and now serving out a seven year apprenticeship at the colliery, saw the Wylam Dilly (and perhaps also Lady Mary) projects through to completion. Apart from the Timothy Hackworth falling-out, there always seems to have been a bad atmosphere in the Wylam camp. As we know, Hackworth's colleague Jonathon Forster clandestinely colluded with George Stephenson, who was working for a rival colliery, which hardly suggests great team spirit. Add this to the fact that Hedley conveyed instructions to his engineers by chalking them on the smithy door, rather than discussing them directly, suggesting a formal barrier between manager and staff. Hedley was always Blackett's man. Happy to adapt his locomotives to break a workers' strike. One of the later pictures of Wylam Dilly shows two of Hedley's sons dressed in frock coats and top hats beside the engine, which is manned by black faced cloth-capped labourers. This says all we need to know about the relationship between the Hedleys and their employees.

After finishing work on 'Lady Mary', Hedley's locomotive building days effectively ended. In 1826 he leased Black Callerton Colliery, to the north of Newburn, and left Wylam for good. He nonetheless maintained a passing interest in the development of steam locomotion, though the pioneering days were over. He bought shares in a number of collieries in South Durham and became involved in the ongoing struggle between rival railway companies in finding the cheapest route to the sea for coal from Central Durham. He threw his weight behind Christopher Tennant's proposal for a railway direct from the Durham collieries to the River Tees and gave Parliamentary evidence in support of what became the Clarence Railway (CR), no doubt to the annoyance of the Stockton and Darlington Railway who had George Stephenson as their engineer/surveyor. Hedley persuaded his former Walbottle buddy, Robert Hawthorn, to build some of the first engines for the CR, updating his Wylam design. These engines were the freight 0-6-0s 'Tyneside' and (inevitably) 'Wylam'. In 1837 he retired to a new home, Burnhopeside Hall, near Lanchester, where he lived until his death in 1843. Despite his faults, he deserves to be recognised as one of the first great railway pioneers.

*The Puffing Billy replica working at Beamish Industrial Museum*

The best tribute that one can pay to Hedley's engines was their reliability at a time when steam locomotion was considered an expensive and experimental folly. Two of the four locomotives he built were robust enough to work for half a century and are still to be seen at the Science Museum and the National Museum of Scotland. It is a measure of their success that, despite the numerous locomotives built by others over the decades following the Wylam engines' construction, the basic design of Puffing Billy was still the one chosen in 1828, by Foster and Rastick at Stourbridge, for the first steam locomotives exported to America. By the mid-twenties, however, things had moved on. The last successful Wylam design engine built was the 'Agenoria', now in the National Railway Museum in York. It was manufactured in 1929 for the Shutt End Colliery in Staffordshire and worked there, in the manner of Hedley's earlier prototypes, quietly and efficiently for 35 years. The year it emerged from the manufacturers however an altogether different type of locomotive was making a dramatic debut. It was built by George Stephenson's son at Newcastle and would henceforth dictate the direction locomotive engineering was to take. It was the 'Rocket'.

# Chapter 17

## *Timothy Hackworth's Archive*

*Portrait of Timothy Hackworth*

On the 17th October 1829, a member of the Stockton and Darlington Railway (S&DR) Committee, William Kitching, wrote to Timothy Hackworth consoling him on losing out to George Stephenson in the recent Rainhill locomotive trials. Kitching was critical of Stephenson generally but especially scathing about Stephenson's engines, which at that time had a monopoly on the S&DR. In particular, he was annoyed about a recent S&DR purchase from Robert Stephenson's Forth Street works at Newcastle which…

'was at work scarcely a week before it was completely comdemd (sic) and not fit to be used in its present state, the hand gearing and valves have no control…'

If Kitching's spelling left much to be desired his sentiments were nevertheless all too clear, Stephenson's engines just weren't good enough. That year, Timothy Hackworth's spirits had been particularly low. He believed himself to be working in a place he didn't want to be and responsible for locomotives that didn't work. The problem was that the engines in question had been given the seal of approval by the most famous locomotive engineer in the world.

*Locomotion on the opening day of the Stockton and Darlington Railway, September 1825*

Hackworth was taken on, at George Stephenson's instigation, by the S&DR with the specific task of maintaining the engines; both stationary and moving, produced at Stephenson's Tyneside factory, in turn partly owned by Edward Pease the Railway's founding father. However, Hackworth could see the flaws in the designs the moment the drawings appeared on his desk. He'd been there, done that and bought the oil stained sweatshirt more than a decade earlier. He had even worked on 'Locomotion', the template for the steam locomotive fleet, when he was seconded by his then employer to assist at Stephenson's Forth Street works, prior to his appointment by the S&DR. The problem for him now was that he had no remit to make changes. He was forced to accept Stephenson's flawed machines as they were, then shore them up the best he could when they went wrong, which they did with regularity. He desperately needed a free hand.

These insights into the early years of railways have come to light following the public release of the Hackworth Family Archive by the National Railway Museum at York. It seems that, by the beginning of 1828, Stephenson had taken his eye off the ball in the north east. His enthusiastic presence and expertise was now seriously diluted by unfolding events at Liverpool, where he had been appointed the engineer for the much bigger Liverpool and Manchester Railway project. Consequently, locomotives supplied by his Newcastle company to Teesside, were not up to mark; indeed little better than the colliery engines he had worked on a decade earlier at Killingworth colliery. One arresting indication of the state of Stephenson's engines is related in Kitching's letter and concerns a hair-raising incident involving Stephenson's brother James (Jem), who it seems:

*'when standing without the wagons at Tullys a few days ago, it (the locomotive) started by itself when the steam was shut of and all that Jem Stephenson could do was not stop it, run down the branch with such speed that old Jim was crying out for help everyone expect(ing) to see them both dashd to atoms the depots being quite clear of waggons which would have been case had not the teamers and others thrown blocks &c in the way and fortuna(te)ly 'threw it off'*

One can sympathise with Jem who it appears was only prevented from being ' *dashd to atoms*' by the prompt action of fellow workers who threw stone sleeper blocks into the path of the locomotive to derail it.

It isn't clear in the letter which of Stephenson's locomotives Kitching is referring to here, as it is only given the telling pseudonym 'maniac', but the incident related was, it seems, not a one-off[21]. The same engine went on another bender the following day, careering out of control and causing panic on the coal staiths at Stockton. This was due to a failure in the primitive steam pressure regulator. It would not be

---

21 The likelihood is it was a 0-4-0 engine called 'Rocket' – not the famous Rocket of Rainhill but an earlier incarnation of the name.

the only serious accident involving boiler pressure control on one of Stephenson's engines. On another occasion, with the unlucky Jem again at the tiller, the boiler of an engine exploded, catapulting the fireman, John Gillespie, 24 yards through the air to die later from the burns he received. Three times that year the lead safety plug in the boiler of the same locomotive melted; the plug being the ultimate failsafe prior to boiler detonation.

All this was especially galling to Hackworth. He knew perfectly well how to build safe and reliable locomotives; at least two of his engines were working quietly and efficiently at Wylam, using positive innovations which were at least partly to his own design.

One of these was the 'U' shaped flue, which recovered more heat from the fire before the smoke was allowed to escape from the chimney. The other was the primitive blast pipe. As a result of their reliability, Puffing Billy and its Wylam successors had gained an international reputation at a time when steam locomotives were still thought of as expensive and unnecessary experiments.

Since the Pease family was George Stephenson's biggest advocate at the time, not to mention part owners of Forth Street, George Stephenson's word, unfortunately, at that time was law; so if Hackworth had reservations over the Forth Street engines there seemed little he could do about it. The first of the S&DR engines was Locomotion or 'Active' as it was originally called. Active it certainly wasn't. It certainly didn't exhibit the *'proved efficiency'* attributed to it by Samuel Smiles in his biography of Stephenson. In the case of 'Active', the wooden spoked wheels literally fell off as may be gathered from an 1826 letter from Stephenson to Hackworth in the Archive:

*'How does the new plan of the wheels do? is there any appearance of working loose? How does the old Engine get on?*

*I hope by this time you have got the Shops covered in; so as to get the Engines under cover to repair them.*

*I hope Robert Morry will get his heart up and try to do better it certainly was a sad mistake in him to leave the Engine as he did at Darlington last Sataurday night.'*

Worse was to come. On the 1st July 1828, Locomotion or Active's boiler exploded, killing the driver John Cree and seriously injuring the fireman, the latest in a series of disasters befalling the same engine. This was a worrying time for Stephenson's employers. In a letter from the S&DR engineer John Dixon, Stephenson, who in Dixon's words was *'once deified'* by the Pease family, had now fallen *'out of favour'*. The Darlington Quakers, who were, less than two years into the operation phase of their public railway, now considering replacing all their once-prized steam locomotives with horses. By mid-summer, there were no serviceable engines left and the proprietors, whatever their commitment to steam had originally been, had no choice but to fall back on horsepower. The best Stephenson could offer were his dandy carts, and even the ones he gave the S&DR were originally intended for the Canterbury and Whitstable Railway (C&WR) and were only transferred to the S&DR following the C&WR's (ironic given the circumstances) decision to rely solely on steam power.

This presented Robert Stephenson with a dilemma. He knew the Liverpool and Manchester Railway company directors wanted to see efficient engines at work before agreeing to their use and, unfortunately, the existing Stephenson built Shildon engines were a poor advert. With a visit to the north east imminent, Robert wrote to Hackworth begging him to do all in his power to get at least some of the S&DR locomotives working. Robert's letter is interesting, not just for what it conveys about the ongoing problems at Darlington but also the manner in which he says it. Compare Robert's elegant text with the earlier letter to Hackworth, from his less educated father:

*'The reports of the Engineers who visited the North to ascertain the relative merits of the two systems of Steam machinery now employed on Railways, have come to conclusions in favour of Stationary Engines. They have increased the performance of Fixed Engines beyond what practice will bear out, and I regret to say they have depreciated the Locomotive Engines below what experience has taught us. I will not say whether these results have arisen from prejudice or want of information or practice on the subject. This is not a point which I will presume to discuss. I write now to obtain answers to some questions, on which I think they have not given full information. Some of their calculations are also at variance with experiments that have come under your daily observation.'*

Both letters, in their own way, demonstrate that the future of steam locomotion was far from secure at the time. The obvious solution was a reliable locomotive that could be paraded in front of the visiting Liverpool delegation. Luckily Hackworth came up with the answer. A year earlier he had obtained his employers' permission to build an engine entirely to his own design, using all the skills and expertise acquired in his twenty years in the business and the result was the 0-6-0 'Royal George'; the King's name perhaps ironic given its association with currently 'out of favour' Stephenson Snr.

*Hackworth freight 0-6-0 'Royal George'*

For 'Royal George', Hackworth used all the lessons he learnt at Wylam. The weight of the boiler was distributed over six wheels instead of the Stephenson norm of four, and the engine was fitted with a return-flue boiler so it didn't lose steam pressure at critical moments. It also had the first true blast pipe, which Hackworth designed and patented, an improvement on the primitive device employed at Wylam. He tapered the steam outlet pipe into the chimney which further increased steam ejection speed and turned the pipe at right angles so it aligned with the flow of hot gases.

Given the improvements, during trials Royal George performed little better than the Stephenson engines. It broke down within two weeks of entering service on the 29th November 1827 and seems to have been inoperable the following summer at the time the Pease family were considering abandoning steam locomotion altogether. Nevertheless, these turned out to be teething troubles and it went on to become the most reliable of the S&DR's engines over the following decade. So steam locomotion survived and flourished on Teesside and Royal George was inevitably the locomotive that was trotted out for the benefit of the visiting L&MR directors.

*Hackworth's 0-4-0 'Sanspareil' at the National Railway Museum, Shildon*

Because of the existing S&DR's financial problems, Royal George was built on a shoestring. Its boiler was recycled from an engine known on the S&DR as the 'Chittaprat', which the S&DR had taken on trial from the firm of Robert Wilson & Co at Newcastle[22]. Chittaprat in its original form hadn't lasted long. It was scrapped during the test period following collision with another engine. However, its boiler, lengthened for Timothy's engine, survived much longer. Royal George included a number of unique features. Apart from being the first locomotive built with six coupled wheels, it had a heat exchanger to recover heat from exhaust steam to pre-heat water fed to the boiler and its weight was supported on metal leaf springs designed to accommodate the variable quality of the rails. It also had, for the first time, a reliable and easily adjustable pressure relief valve; a relief in itself to S&DR drivers given the unsavoury record of Stephenson's engines.

From then on Hackworth and Stephenson were as much competitors as colleagues. For the Liverpool and Manchester Railway (L&MR) locomotive trials at Rainhill, for example, Hackworth designed and built his own locomotive, Sanspareil, which competed with Stephenson's Rocket for the £500 prize. As we know Rocket won, in part because of the efficiency of its multi-tube boiler, which outperformed the Wylam derived return flue adopted for Hackworth's engine. As it happened, Hackworth couldn't really complain about the outcome as Sanspareil should never have been allowed to compete in the first place, since it exceeded the prescribed engine weight. Despite this, Sanspareil's defeat at Rainhill left a sour taste in the mouth of Timothy Hackworth who howled 'conspiracy'. Unfortunately, he had only been allowed by his employer to work on Sanspareil in his own time with the result that the locomotive was hastily assembled. He was also given no opportunity to test it out prior to the trials. Worst of all, time constraints meant he had to subcontract manufacture of certain key components, rather than engineer them himself, and one of these, a piston, leaked badly and then broke during one of the engine's demonstration runs. The piston, as it turned out, had been manufactured at Forth Street, fuelling forever the theory that the Stephenson family conspired to rob him of the Rainhill prize. Wounded by the experience, Hackworth wrote to the L&MR Directors begging them to reconsider their verdict; barely disguising his belief he'd been cheated:

---

22  Chittaprat was an onomatopoeic name derived from the sound the loco made.

*'You are doubtless aware that on a recent occasion the Locomotive Engine Sanpareil failed in performing the task assigned to her by the judges – it were now useless to enter into a minute detail of the causes – suffice it to say that neither in material construction nor in principle was the engine deficient, but circumstances over which I could have any control from my peculiar situation compelled me to put that confidence in others which I found in sorrow was too implicitly placed – as the defects were of a nature easily to be remedied – my immediate attention was turned to that point – and I now report to you the extent to which success has attended my efforts. The whole alteration which has been made is the removal of a cylinder which failed from its defective casting.'*

In Hackworth's view, it was only the defective Stephenson cylinder that let the engine down, otherwise his locomotive was sound and should have taken the prize. He considered that his confidence in Stephenson, had been *'too implicitly placed'*. The fact that the L&MR offered to buy Sanspareil for £550 did little to soften the hurt caused by *'circumstances over which* (he) *had no control'*. Even eighty years later his grandsons, Samuel Holmes and biographer Robert Young refused to accept that Timothy had been beaten fair and square. Young wrote:

*'Sanspareil's breakdown was entirely owing to the bad workmanship of Robert Stephenson & Co., who cast and bore the cylinders, the very firm which was competing against (Hackworth) and to whom he had perforce to go.'*

Despite Rainhill, Hackworth's confidence in the abilities of Sanspareil (and Royal George) was not misplaced. Both engines enjoyed a long and useful working life. It is a testament to Hackworth's expertise as an engineer that, after Royal George, virtually every freight engine used on the S&DR was built to Hackworth's design, including those manufactured by Robert Stephenson & Co. In total, eleven new engines were produced between 1827 and 1829, using lessons learnt from Royal George, including the 'designed for speed' Globe; the first locomotive used on the S&DR solely for passenger traffic.

Hackworth had moved to the S&DR locomotive works at Shildon in 1826 and worked there for fourteen years. Heavy freight engines were his speciality. He reluctantly discarded the Wylam 'U' tube flue in preference for Stephenson's more efficient multi-tube boilers. The next generation of Shildon engines, which benefited from this modification, were known as 'Wilberforce' engines and, according to the S&DR, were the best ever used on the line. He terminated his contract with the Pease family in 1840 and took over the running of the locomotive works a mile or so to the east. This had, since its creation, been managed by his brother, Thomas, but the works was only there at Timothy's instigation, and was (at least partly) financed by him. Hackworth's brother Thomas had also served his apprenticeship at Wylam and was familiar enough with the Wylam locos to later add his own postscript to the 'Who invented the blast pipe' debate, unsurprisingly crediting Timothy as the inventor. The Hackworth owned factory was known as Soho Works and Timothy worked there for the rest of his life, producing many more steam locomotives, including a fleet of engines for the London Brighton and South Coast Railway.

He even managed to design the first freight locomotives used in Russia and Canada.

The Hackworth Family Archive reveals Timothy to be first and foremost a family man. He fathered nine children, three sons and six daughters, although one of his sons died in infancy, and much of the correspondence in the Hackworth Archive at the NRM relates to family matters, in particular the anguish Timothy felt over his daughter Ann's mental illness. Nevertheless, throughout his life, his religious faith never wavered. He continued to act as a Methodist lay preacher right to the end of his life, conducting services all across County Durham, including regular Sunday sessions in Barnard Castle. After his death in 1850, his son John Wesley Hackworth took over the works, but by then the railway boom had run its course and locomotive manufacturing at Shildon was in decline. Soho works was sold in 1855. The land on which it stood, along with some of the buildings there in Hackworth's day, are all now part of the National Railway Museum. John Hackworth became his father's main advocate, voicing Timothy's claim to be the 'father of the steam locomotive' to anyone who cared to listen. The Family Archive contains many similar letters from John, written on behalf of his father to newspapers and technical journals. He was particularly affronted by the, unquestionably biased, biography of Stephenson written by Samuel Smiles, which relegated Timothy to a bit player in railway history. In a typical undated diatribe, to the Northern Daily Express, written sometime around 1858, John argued that not only was Timothy Hackworth, and Timothy Hackworth alone, the inventor of the locomotive blast pipe but that most of Samuel Smiles references to his father in Stephenson's biography were fictitious. He did a similar hatchet job on the biography of William Hedley written by Hedley's son Oswald, that credited William Hedley, the Wylam viewer, and not Timothy Hackworth, as the engineering pioneer at Wylam. Hedley, said John:

*'was no mechanic, neither did he invent any part of the locomotives employed there…..'*

Timothy's later years were troubled by illness but his enthusiasm for steam never wavered. Not long before he died, he wrote in a letter to his son John that touchingly ends *'pray for me amen'*:

*'I believe myself that something more will be done, the Locomotive Engine is capable of vast improvement; …my own opinion is that something different from anything yet invented will be brought bear, in a much simpler manner than anything yet in use – simplicity & economy must be the order of the day.'*

There could be no better epitaph.

# Chapter 18

## *John Hackworth's 'Tsar Trek'*

*John Wesley Hackworth*

Buried in the wealth of material that comprises the Hackworth Family Archive is a strange poem. It is unsigned, undated and badly written on note paper headed 'Priestgate Engine Works, Darlington'. It is titled, *'On the Return of Our Friends from Russia'* and concerns the creation of the first conventional railway in Russia and the extraordinary meeting between Russian Tsar Nicholas I and a small party of railway technicians from England, including John Wesley Hackworth, aged 16, Timothy Hackworth's eldest son.

It is fair to say that Russia was a late starter in respect of railway construction. Despite a desperate need for communication between urban centres in such a vast country, by 1836 transport had advanced little since the middle-ages. In some ways this is surprising; as a 20 year old, Grand Duke Nicholas had seen the future of transport firsthand. During the sort of education provided for a future Tsar, he had been despatched on a pan-European 'Grand Tour', spending several months in England and travelling widely throughout the UK. His interest, it must be said, had more to do with finding better ways of killing his subjects than learning about the future of transport. Always more interested in ordnance than industrial technology it must have come as quite a shock to the operators of Middleton Colliery in Leeds when he turned up out of the blue to cast an imperious eye over their novel steam railway.

*Tsar Nicholas I*

In 1816, when this took place, the Middleton Railway was already well-established. It was built by a Geordie, John Blenkisopp and designed on the rack and pinion principle. The locomotives were engineered by Blenkinsopp himself and, like most of his contemporaries, Blenkinsopp thought locomotive wheels wouldn't grip without some additional aid to keep them on the rails. Middleton Railway, which ran from a local colliery to coal drops on the River Ayre in Leeds, was therefore constructed on the same principle as mountain railways such as that on Snowdon today. As a consequence of this, the engines used at Leeds, if reliable, were not meant for speed, indeed they offered few advantages in terms of overall transport costs over horse-drawn wagons. What Nicholas thought of the Middleton Railway after his visit is not recorded but it seems likely he also joined the throng of foreign visitors to Wylam and Killingworth where conventionally wheeled locomotives were also working on a daily basis. On his return to the Motherland his father, Alexander I, commissioned a survey of all the railways in north east England. That a railway system for his country wasn't considered a priority may be inferred from the fact this survey was still incomplete at the time of Alexander's death in 1824. To commemorate Nicholas's visit to Leeds, Blenkisopp provided a working model of one of his locomotives to take home.

If impressed by what he saw, Nicholas displayed little interest in railways in the years immediately following his accession to the throne in 1825. This may have been because he had other important matters to deal with, including a number of attempts on his life. So it was only when he had crushed the opposition that he was able to turn his attention to a transport system designed to connect the widely disparate cities of his vast empire. No mean feat, it is nearly 4000 miles between Moscow and Valdivostok.

By now, Russia was thirty years behind the rest of the world in terms of railway technology and Nicholas thought that every element of railway infrastructure, from track to locomotive, would have to be expensively imported. This wasn't entirely the case. In actuality, there were already a small number of private railways operating throughout his empire, although nearly all these used horses for haulage. However, in the far-away Ural Mountains, there was in fact a working steam railway worthy of mention.

By the end of the eighteenth century, the mechanics of stationary steam engines were well understood and so, as in England, it was inevitable that those with sufficient engineering ability to construct a Newcomen or Watt steam pump would eventually turn their attention to applying the same technology to turn the wheels of vehicles . However, in a country so vast, and communication so poor, there was little opportunity for sharing knowledge so it wasn't until thirty years later that the first steam locomotives were built for a copper mine in the Urals run by the father and son team Cherepanov. After dismantling and re-assembling various types of stationary engine and, following a visit to England in 1830, the Cherepanovs decided to have a go at building their own locos. Their intention was to use them for moving copper ore from their mine workings at Demidov to a nearby foundry. There followed the usual Wylam style trials and tribulations. The first loco burst its boiler during trials and had to be completely rebuilt. Made from wood and copper with a small iron boiler it was not a success. The second engine, like that at Wylam, fared much better. It was provided with a larger boiler and maintained a decent steam pressure, successfully completing trials early in 1834. The original engine was soon modified in the same way.

A short length of 5.5 feet gauge railed track was laid within the factory precincts and a wagon jam-packed with forty passengers was attached to the loco and the train paraded back and forth before an assembled crowd of employees, reaching a top speed of 10 mph. Both engines superficially resembled Stephenson's Rocket, which was seen by Cherepanov Junior during a visit to England in 1830, nevertheless much of the design was unique to the Russian engines because, at the time of his visit, the young man spoke no English and the railway people he met spoke no Russian, so he learnt little about the mechanics of the machines. In the summer of 1835, both locos were in daily service on the 1.5 miles of track between the factory at Vyskii and the mines at Mednyi. However, by this time, hundreds of miles away in St. Petersburg, work had begun on the first Russian passenger railway.

Having (literally on occasions) killed off the opposition, Tsar Nicholas now felt sufficiently secure to start thinking about a national transport system. A network of canals had already been established for bulk movement of non-perishable goods and minerals but was considered unsuitable for food transportation or any other use where speed was essential. If Nicholas had heard of the Cherepanov railway he must have discounted it because, ignoring the objections of his own ministers, he decided that the technology needed would be imported – and preferably from England.

As the recognised world leaders in the technology, British railway expertise was much in demand, with the locomotive factories of Stephenson in Newcastle and Hackworth in Shildon often the first port of call for foreign visitors. Timothy Hackworth, after several years working for the Pease family, had also set up his own business. Orders were piling up for engines from his works in Shildon when Tsar Nicholas reappeared on the scene.

Nicholas was only a recent railway convert. Up until then he had been unconvinced about the merits of railways, his reticence perhaps understandable given the severity of Russian winters compared with those in England. He is also said to have *'seen railways in England and found they were dangerous and ran over people'*. Whether he had personally seen anyone mowed down is improbable but he may have discussed the matter with one of his personal guards who was present at the opening of the Liverpool and Manchester Railway and could well have witnessed the demise of William Huskisson M.P. under the wheels of Stephenson's Rocket.

The need for fast and efficient transport in Russia was imperative. Nicholas needed a railway, if only for defence purposes. It was decided, in the first instance, to construct a railway from the Baltic port of St. Petersburg to the capital at Moscow. The money needed for construction would be raised partly from shares and partly from the Tsar's own coffers, leaving an option for the railway company to buy out the Tsar's interest later. A Czechoslovakian engineer called Franz von Gerstner was given the task of overseeing the project. The railway was to be engineered in short sections, each section independently financed, and the first 15 miles of single track would be laid down between the Baltic port of St. Petersburg and Tsarkoye Selo, where, unsurprisingly, the Tsar also had one of his summer palaces. Seven locomotives were purchased from abroad, six from England and the other from Belgium. The original intention had been to purchase four of the British engines from Robert Stephenson's works at Newcastle but, due to other commitments, Stephenson was only to supply two directly, with Hackworth's Soho Works at Shildon subcontracted to provide the remainder (the other two UK engines were purchased from the Vulcan Foundry at Newton-Le-Willows).

The order books at Hackworth's Soho Works were also full to overflowing and Timothy's factory would consequently be stretched to the limit to accommodate the work. After construction, the loco would be broken down and shipped in crates to Russia. Although the obvious candidate to make the journey, Timothy's brother Thomas, who had helped to build the engine, couldn't be spared and so Hackworth's young son John was given the formidable responsibility of overseeing the transport of the Shildon loco.

John Wesley Hackworth was born at Walbottle near Newcastle in 1820. His earliest memory was of the Stephenson's locomotive building works at Forth Street where his father was acting manager. He wrote of his years on the S&DR:

*'I saw it opened, was brought up upon it, knew every horse and driver, every loco driver, every director, nearly all the shareholders, every noteworthy incident that occurred thereon for the first twenty years.'*

By the middle of the third decade of the century he was serving an apprenticeship in Soho Works. As a youth he was tall and heavily built and even then commanded respect from the workforce, so it raised few eyebrows amongst his work colleagues when he was tasked with overseeing the transport of the locomotive to St. Petersburg. His father knew it would greatly enhance the reputation of his company if it was Shildon which provided the first working steam locomotive (or so it was generally believed) used in Russia and all haste was therefore made to complete the construction before the Stephenson built engines could emerge from the Newcastle workshop. The Shildon loco had a six foot wheel gauge, in accordance with Von Gerstner's instructions, although it later transpired that the St. Petersburg railway would be the only part of the Russian railway system to use that particular gauge.

As today, access to Russia for outsiders wasn't straightforward. When John applied for a visa his application was refused and it took the personal intervention of the Tsar to secure him entry to the country. Even so, one of his Christian names wasn't considered suitable for a visitor to the Mother Country, the Hackworths you see were Methodists. Even worse, Timothy was a Methodist lay preacher, and Methodism with its overtones of universal suffrage and education for all regardless of wealth wasn't quite the thing the Russian Orthodox Church wanted to encourage. In consequence, John Wesley Hackworth became John William Hackworth for the duration of his visit.

The Priestgate poem is scathing about the first meeting with the Tsar:

> *'The grip of his hand did it not make you blush,*
> *If it didn't your soul is as dull as ditch water,*
> *For of murderers by wholesale sure Nicholas is one,*
> *Only think of the cold blooded slaughter,*
> *And the wretch singing psalms as the massacre done.'*

By autumn, the construction of the first Hackworth loco was complete. It was built in accordance with Stephenson's 2-2-2 'Patentee' design, as recently used to effect on the Liverpool and Manchester Railway; the Stephenson design confirming that Timothy Hackworth was only acting as Robert Stephenson's sub-contractor here. The engine was tested at Shildon on a roller-bed where it reputedly achieved a top speed of 72 mph, 10mph faster than its Newcastle-built sisters. Shipment of the locomotive was arranged from Stockton-on-Tees, where the engine was delivered on S&DR rails and loaded on to the brig 'Barbara', which set sail on the 3rd October 1836 accompanied by the small support party led by John Hackworth.

Cold as it gets in the north east, John wasn't prepared for the sort of weather common to Eastern Europe in late September. Not long after the boat entered the Baltic it founded in pack ice. The dock at St. Petersburg, where the ship had intended to berth, was icebound and the ship and its precious cargo was forced to retreat and seek refuge in the first available ice-free port. There, arrangements were hastily made to purchase horse-drawn sleds, which were loaded with the engine's components parts for transport overland to Tsarkoye Selo. At the end of the first day, the men made camp overnight in the snow and the smell of cooking attracted a pack of hungry wolves which surrounded the camp and attacked once it got dark, not the sort of thing John would have been prepared for during his apprenticeship in England. An all-night vigil had to be maintained to protect the horses before the wolves eventually slunk away at first light and allowed the convoy to continue. The following day they reached Tsarkoye Selo where, to John's dismay, it was discovered that vital repairs to the engine were now needed. Some of the metal components had suffered in the sub-zero temperatures and one of the cylinders had cracked. Hackworth's Shildon foreman, George ('Geordie') Thompson was sent off to Moscow on horseback, a distance of 600 miles, where there was the nearest ordnance factory capable of mending the broken part. Even acknowledging his Olympian gallop through the snow, Thompson also showed much initiative. He arranged for a mould to be made, oversaw the casting and boring of the cylinder, and then rode all the way back to St. Petersburg and fitted it on the engine.

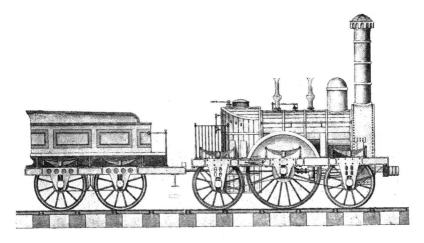

*Hackworth's 2-2-2 Russian engine*

While this was going on, a short length of line was in the process of being laid in the grounds of Tsar Nicholas's summer palace where, on the 3rd November, the engine was trialled for the first time. Considering the earlier problems, the trials went well and Nicholas gave the locomotive the Royal thumbs up; complimenting John on the technological advancements made since his earlier visit to England.

The engine was then moved to the new Tsarkoye Selo railway station, where it was formally blessed according to the arcane requirements of the Greek Orthodox Church. This amounted to encircling the engine in a ring of holy candles and then baptising it before an assembled congregation. The 'holy water' used had earlier been collected from a nearby marsh and rendered 'holy' by immersing a gold cross in the bucket in which it was being carried. While a choir chorused the blessing, each of the locomotive's wheels was anointed individually by a priest, before the audience of assembled dignitaries, which included a bemused John Hackworth. All onlookers were then invited to anoint the rest of the machine with holy water, following which five coaches were hooked up and, loaded to overflowing with passengers, the train set off. Reaching a top speed of 12mph it steamed south to Pavlosk, where short excursions in a wagon were arranged for anyone wishing to ride behind the engine. Tsar Nicholas's view of the Hackworth engine was, *"It is the finest that I ever saw."*

We can, however, take this with a large pinch of Siberian salt since there is no evidence that the second Hackworth locomotive, completed the following year, was ever delivered to Russia[23].

On completion of the ceremony, the Tsar was formally introduced to all the Englishmen, including one young man described as a 'tanner'. According to the Priestgate ode, young Harry the tanner briefly swapped his beaver-skin cap for the monarch's crown:

*'O Harry it's true that Nic tried on your beaver,*
*And placed on your head his own helmet of brass,*
*And then nearly laughed himself into a fever*
*He looked such a comical sight in the glass.'*

The trials were not, it should be said, a complete success. It had been anticipated that birch wood would provide the fuel, since it was cheaper than coal and readily available. However, it caused a lot of sparks which fell out of the chimney on to passengers in open carriages and set fire to their clothes. Although the burning apparel were quickly doused this can only have dampened the passengers enthusiasm for the event somewhat.

By and large, however, the trials went well, as did the outings of the other locomotives which arrived early in the New Year. To

---

23  In fact, the second engine was almost certainly purchased by the S&DR and, after a less than successful trial period, was used solely on passenger services on that railway.

commemorate events, a medal was struck with the image of Nicholas I on one side and a drawing of a Stephenson 'patentee' engine on the reverse. Still sensitive to his global reputation, Nicholas withdrew the medal from circulation when he discovered that the inscription on the reverse side of the medal described Von Gerstner, a Czech national, as the architect of the railway, rather than himself. Nicholas was still unconvinced about the advantages bestowed by railways. In consequence, it would be another ten years before St. Petersburg and Moscow were connected by rail. At the out-break of the Crimean War in 1852, Russia still only had 650 miles of railway. Only when confronted with the desperate communication problems manifest in the conflict, did the Tsar reconsider and institute a major programme of railway construction, which he funded by selling the apparently resource free wilderness of Alaska to the United States – a decision for which the U.S. oil industry is forever grateful.

John Wesley Hackworth returned to England to support his father at Soho Works but, as the outward manifestation of his new-found maturity, he now sported a beard, unfashionable at the time in England, which he retained for the rest of his life. Over the following years, the Shildon locomotive business would fall into decline; the Hackworth designed engines looking increasingly dated with each passing year. John's final attempt to revive interest in his father's business was the construction of an engine called 'Sanspareil 2'. This loco, patented jointly with his dad, was intended to be the last word in locomotive design. It was, however, the last throw of the dice for Soho Works. John touted it around the country's numerous railway companies but generated little interest. He even issued a 'Rainhill' type loco-versus-loco contest to Robert Stephenson; challenging him to compete under any terms he wished, with Sanspareil 2 pitted against the best Stephenson's company could offer. Stephenson declined and John was left with an expensive engine on his hands he couldn't sell. His interest in railways might be said to have ended there. Soho Works soon passed into the hands of the Pease family who sold it off piecemeal. John became more bitter and angry as the years went by. On his father's death in 1850, a family row broke out between John and his younger brother, also called Timothy, over ownership of his father's works. John wanted to sell the business and accompanying land whilst his brother hoped to save it. In truth, it was John who was right; the works were in a poor state. The machinery and infrastructure was decrepit and outdated and needed a massive influx of money to restore it to anything resembling its former glory. In the end, a modest value of £5000 was placed on land, works and cottages, including his father's home.

After a protracted three year battle between the Hackworth brothers, the works was finally sold. John used his share of the proceeds to set up an engineering works in Darlington but, disillusioned with railways, he abandoned locomotive manufacture in favour of the production of industrial machinery, specialising in his own patented horizontal high pressure steam driven lathes for the cotton industry. He spent the years leading up to his death in 1891 slagging off contenders to his father's right to be heralded as the world's greatest locomotive engineer. Consequently, whenever a new biography of George Stephenson, Richard Trevithick or even lesser lights such as William Hedley appeared, John was there challenging the content. There are no biographies dedicated to John Wesley Hackworth. He had had, nevertheless, his one great moment. To finish with another verse from the Priestgate ode:

> *'most of the sons hitherto have not turned out great men,*
> *But now there's a hero who bearded the Czar in his den'*

**Notes to Chapter 18**
Despite terminating his career as the builder of steam locomotives he continued to have success as an engineer from his own Priestgate works at Darlington, manufacturing stationary engines and other industrial machinery. He was responsible for generating several notable patents, including one for a 'dynamic valve gear, which dramatically improved the efficiency of steam engines.

# Chapter 19

## *William Churchman and the 1911 Rail Strike*

*The workforce at Shildon*

On the 18th August 1911 members of the four main rail trade unions withdrew their labour and initiated the first national rail strike. Although this was not the first time railway workers had downed tools, it was the first time industrial action involved the whole of the rail network.

Disputes between management and railwaymen, on a local level, had been regular events since the beginning of railways; even in those early years the relationship between management and workers was fractious at best. At Wylam, for example, where Puffing Billy and Wylam Dilly were arguing the case for steam, there were a number of industrial incidents resulting from management/worker disputes. The most serious of these involved actual engine derailments caused by drivers of horse-drawn wagons, whose livelihood was threatened by the appearance of the locomotive. A particularly notable withdrawal of labour occurred in 1815 when senior blacksmith, and de-facto railway engineer, Timothy Hackworth, refused to work on the Sabbath, occasioning both his hasty departure, and clearing the way for his pioneering endeavours on behalf of the Stockton and Darlington Railway (S&DR).

The S&DR itself was also not immune to industrial action. As early as the initial period of railway construction, between 1823 and 1825, company reports show that work was regularly held up because of management/workforce disputes.

All the disputes, however, involved individual companies and rapidly fizzled out, usually to the satisfaction of management. Following the merger of large numbers of small railway companies, and the consequent expansion of the workforce within a particular organisation, the threat of industrial action carried much greater weight. More importantly, the introduction of trade unions raised the possibility of collective action, which strengthened the workers' hand and made it more difficult for employers to resolve disagreements by simply dismissing the disaffected railwaymen involved. Given the obvious advantages of union membership, it seems surprising that the trade union movement took longer to get established on railways than in comparable heavy industry. According to the Oxford economist and historian Harold Pollins, this was because:

*'...the bulk of the labour was unskilled. The skilled men, the footplate crews were in general well looked after by the companies and had little in common with other railway employees.'*

In other words, the company exploited the principle of divide and conquer; technically able and educated men were looked after, the rest could go hang. It was therefore the unskilled majority who came off worst when it came to workplace pay and conditions. The first trade union to take collective action acted on behalf of footplate-men and was called the Engine Drivers and Firemen's United Society (EDFUS). EDFUS objected to the minimum ten hour working day, not least because serious accidents resulted from drivers working long hours. Having negotiated an agreement with most of the prominent railway companies, they reached an impasse with both the North Eastern Railway (NER) and London Brighton and South Coast Railway (LBSCR). Although the LBSCR offered to reduce the working week to sixty hours, they refused to consider shorter work days, which still permitted the employer to require footplate-men to work unlimited hours in any one shift. For their part, the NER agreed to a maximum twelve hour working day but only if the men also worked occasional unspecified (and unpaid) overtime. On being informed of this decision more than 1000 men on the NER came out on the 11th April 1869. The company responded by recruiting blackleg labour, whilst simultaneously instituting criminal action against the strike leaders, citing breach of contract. The Union hadn't enough capital to mount a legal defence and the strike consequently petered out, with the union itself folding shortly afterwards.

By the turn of the century, management/workforce relations had reached at an all-time low and employee walkouts were commonplace. Following one such walkout in 1907, by men of the NER, a Royal Commission was set up to resolve the differences. Their principal recommendation was that, before strike action could be taken by a trade union, the dispute would be referred to an 'independent' arbitration board who would raise the men's grievances with the companies concerned, in the presence of representatives of the government. Crucially missing from these discussions were the unions themselves, who were not invited, despite the fact that the outcome was binding upon them. Like all such agreements, it was going to be a fudge. The railway companies refused point blank to negotiate directly with the trade unions and so it was left to the government appointed Commission to act as honest brokers.

The railwaymen's financial position was neatly summarised in a poem titled 'Eighteen Bob a Week', published in Railway Gazette, which contained the following telling lines directed at the management:

> *'To them of course it is quite just to draw a princely ration*
> *While we must get our grub on trust and slave at some dull station*
> *Be bullied here and bullied there and take it all quite meek*
> *To drive away all earthly care with eighteen bob a week.'*

The day after strike action was initiated Railway Review reported that industrial action was solid across the country, apart from any metropolitan signing-on point other than Southall.

Nevertheless, in those parts of the country where union membership was strongest, strike action was both solid and militant. In South Wales, where the coal miners were also fighting for better pay and conditions, the railways came to a standstill, despite use of blackleg labour. A typical incident was reported by a passenger on a train from London to Cardiff on the evening of the 18th August. The train was prevented from moving out of Landore station by crowds of strike supporters who spilled on to the line, while eleven miles further on, at Llanelli, crossing gates were manned by pickets closing the line altogether. After three hours passed by for the passengers delayed at Landore a troop train, packed with men from the Lancashire Regiment, went by bound for Llanelli. At Llanelli some 3 – 4000 pickets had gathered behind the closed level-crossing gates. The first train to arrive at Llanelli station was boarded by pickets, the crew evicted and the fire extinguished in the engine. The soldiers who soon arrived on the scene claimed to be outnumbered and unable to deal with the situation without serious loss of life. Meanwhile, back at Landore, the train waiting there was finally allowed to proceed. Just beyond Llanelli station it was stopped again by the thousands of men, women and children who supported the picket. The footplate crew were then evicted from their cab and the guards' van taken over by strikers who made political speeches to crowds from the van's open doors. A further light engine which arrived shortly afterwards was dealt with in the same manner. It was only following the appearance of another 250 soldiers from the Worcester and Devon Regiment that the line was eventually cleared. Contemporary accounts noted that most of the signal boxes along the line remained unmanned throughout the day.

This raises the subject of rail safety during the dispute. It seems that passenger safety was a secondary consideration. Many of the blackleg drivers the companies employed had previously failed normal health tests. These, worryingly, included eyesight tests. Since signal boxes were seemingly unmanned during this period, the inability to see signals was not, presumably, considered a problem.

Of all the stories concerning the 1911 strike, none are more fascinating or bizarre than that relating to the confrontation between the Station

Master at Shildon, William Churchman, and a crowd of striking railwaymen and their families. Shildon folk weren't famous for their tolerance and sobriety. As Robert Corkin put it in his book' Shildon – Cradle of the Railways':

*'There is no doubt that alcohol played a major role in the troubles that plagued Shildon. Engine drivers and firemen seemed to spend every penny they possessed on ale, then after 'a belly full' plan a full scale battle with the police or with anyone who tread their path.'*

*The signal box at Shildon where Churchman took refuge*

It was unfortunate therefore that Shildon railwaymen should be drawn into a confrontation with perhaps the least sympathetic official in the United Kingdom. To say that Churchman was unpopular before the strike began is to barely hint at the truth. Born in the heart of rural James Herriott country, at West Rounton near Northallerton, he was drafted into the coal heartland by the North Eastern Railway. From farming stock, he had little sympathy for either coal miners or trade unions and, in particular, those members in his own employ engaged in industrial action. Indeed, he considered it his duty to deliberately go out and antagonise the strikers by encouraging and applauding the arrival of blackleg labour. Things came to a head on 20th August when Churchman encountered a picket outside Shildon station as he arrived for work. He lost his temper and directed a stream of foul abuse at his erstwhile colleagues on the picket line. They responded in kind, chasing him to the far end of the platform where he sought refuge in a signal box.

*Enniskillen Fusiliers deployed at Shildon during the 1911 strike*

Far from then keeping a low profile, as might have been expected given that the mob outside was baying for blood, he stood at the window, slowly devouring and seemingly relishing, his packed lunch; the intention being to demonstrate that, unlike those outside, he was a wage earner and could therefore afford to eat. Responding to the provocation, the men outside attacked the signal box with whatever they could lay their hands on, smashing the windows with ballast from the track-bed. The angry mob spilled out on to the line and into the adjacent marshalling yard where they boarded a coal train, manned by blackleg labour. There, after a short fight, they evicted the driver and fireman and then rendered the locomotive inoperable by extinguishing the fire. Churchman, meanwhile, waited for the coast to clear then legged it home. He lived nearby, in one of the Soho

Cottages that now form part of the National Railway Museum at Shildon. Unfortunately, the mob had anticipated his appearance and was waiting there for him. He was able to get inside and barricade himself in whereupon the outside of his house was given the same treatment as the unfortunate signal box.

The outcome of these shenanigans was that 200 soldiers of the Enniskillen Fusiliers were drafted to Shildon to guard the station, signal boxes and marshalling yard. Despite their presence, the following day strikers were able to board another train about to depart from the marshalling yard and, as before, eject the blackleg crew. With troops concentrated around the station area the strikers moved on to the nearby Brusselton incline where they released a line of coal trucks which careered downhill to collide with wagons in the marshalling yard, scattering the assembled troops like so many pins in a bowling alley.

If the general public were sympathetic to the strikers' case, the press (as today) was hostile. Angry correspondence from company shareholders was gleefully reproduced in the Times and similar broadsheets. Typical of these letters is that from a curate called Mitchell who wrote an indignant open letter to Lord Stallbridge, Chairman of the LNWR. The clergyman, convinced that widespread revolution was imminent, declared:

*'they (the strikers) will from this week's experience resort to more cunning tactics and deadly weapon such as bombs which from hidden coverts they can throw at the military who even with quick-firing guns will be powerless against their violence and mischief.'*

In the same spirit of love of his fellow man he continued:

*'The House of Lords is now impotent to dispense justice, the shareholders can only look therefore to the Directors to be the judges themselves to see that justice is decreed and unflinchingly carried out. How unjust to reinstate these strikers and workers of mischief! And how unjust and tyrannical thing it is that the companies should be forced to do so.'*

In reality the nationwide industrial action petered out after only three days and life on the railways proceeded more or less as before with little if any concessions made to the railwaymen's grievances. Industrial strife, consequently, didn't end there. Within months the unionised workforce of the NER were out again when one of their number was unfairly dismissed after being caught drunk by his manager in his own time on one of his rest days. The advent of the First World War then intervened and management/worker hostilities were postponed as the nation united against a greater enemy. As for William Churchman, he continued as a station master for the NER but was prudently moved far away from Shildon.

**Notes to Chapter 19**

After the war ended hostilities between the management and the staff continued much as before. There was even talk of nationalisation to resolve the discrepancies in pay between staff, doing the same job, from company to company, but, inevitably, the outcome was a fudge. From the legion of small, medium and large railway companies which existed prior to 1914 four new companies emerged which absorbed the interests and infrastructure of their predecessors namely, the London and North Eastern Railway (LNER), Great Western Railway (GWR), Southern Railway (SR) and London Midland and Scottish Railway (LMS). Within three years of the creation of the 'Big Four' there was another general strike.

# Chapter 20

## J.R. Whitbread and the Railway Police

*John Robert Whitbread*

Travelling by train can be a tricky business. In 1846, a certain Mr. Parker was quietly minding his own business in an otherwise empty compartment, on the journey from Derby to London, when the train stopped at a small station and a man entered. There was no corridor on this train so it was just luck as to who Parker ended up with as a travelling companion. Unfortunately, Parker's luck was out. The other occupant, well-dressed and well-spoken, seemed initially to be a *'courteous gentlemanly man'*. However, after exchanging a few pleasantries Parker's fellow passenger got down on his knees and began to pray. Prayers over, he removed all his clothes and, once naked, tried to force his head through the closed carriage window, badly cutting his face and neck in the process. Thwarted in this bizarre enterprise he took out his annoyance on poor Mr. Parker whose screams for help went unanswered. In the ensuing struggle, Parker somehow got the outside door to the compartment open and climbed on to the running board along the side of the carriage. With the train now moving at full speed, he edged his way along the side of the coach and banged on the window of the next compartment screaming to be let in. His birthday-suited fellow passenger followed close behind but with the assistance of other passengers Parker managed to prevent his naked nemesis from getting in. After a lot of banging on the window and shouting, the man gave up the pursuit and leaped off the train into a field where he was captured and taken into custody by a bunch of astonished farm workers.

This is only one of the fascinating tales in John Robert Whitbread's wonderful book 'The Railway Policeman'. As may be inferred, this is less a book about railway crime, more one about railway policeman, even though in the instance cited policemen didn't figure in the story.

Contrary to popular belief, railway police, in some guise, had been around since the day railways started operating. In the early years, however, the policemen were company employees, usually labourers or navvies, given the additional onerous task of maintaining the company's numerous (and often petty) byelaws. On the first public railways, investigating injuries to passengers seems to have been the least of their duties. Life was cheap. When asked by the Earl of March in 1839, on behalf of Parliament, to provide complete details of all fatal accidents on the Stockton and Darlington Railway (S&DR) the company could only manage a few hastily written and barely legible notes. One typical example is this terse note drawn from the company records of 1826:

*'June 30th Mr.Glass's son killed near Ellerby.'*

This is in fact one of the wordier entries in the list, so presumably 'Mr. Glass' was someone well known to the S&DR. Often the only comment against any incident report was the date of the event and a couple of words, for example, 'Child killed'. Deaths on the nascent railways were commonplace. Amongst the listed fatalities in the report to parliament of the S&DR is that of Robert Stephenson. Obviously this wasn't George's son but could well have been one of George's wider family, who were all employed in some capacity on the railway. This particular Robert Stephenson was killed by moving trucks after 'leaping from a stone waggon' in order to try and remove a log blocking the line.

Amongst their duties on the S&DR, company policemen were expected to investigate all 'accidents on or to any part of the railway'. The records do not enlighten us as to whether this included the death of one particular man who died from *'alcohol poisoning'*, having illegally tapped into a keg of spirits on the journey between Stockton and Darlington. Drunkenness, it appears, was a common factor in many of the early deaths. It certainly contributed to the demise of John Philips, publican of Yarm, who in 1839 *'eluded the railway police in the dark'* and was run down by a train near Urlay Nook as he wandered along the rails *'in an advanced state of intoxication'*. Despite the inherent danger of the work, railway policemen were not the most highly paid of employees. By the mid-1830s their rate of pay was about a pound per week; the same as porters, and less than signalmen ('switchmen' before the days of signals), locomotive drivers and firemen. For this a lot was expected. The S&DR for example, in 1835, employed just one superintendent and four regular officers to cover their, by then, extensive network. These men worked every day of the week apart from Sunday from 6am to 8 pm with four hours additional work expected on the sabbath. They had much to do. In addition to accident investigation they were responsible for preventing trespass, reporting excessive train speeds, prevention of any criminal activity in whatsoever form it might take, coupled with the investigation of the misdemeanours of company officials, particularly engine drivers who had the disturbing habit of halting trains to take on liquid refreshment at the many public houses which arose trackside.

Railway policemen sometimes also acted as signalmen. On the London and Birmingham Railway (L&BR) constables were installed in sentry boxes, located 1 to 1.5 miles apart along the entire ninety mile length of line. The main responsibility of these switchmen was to regulate traffic movement and the methods used to control train movement varied from company to company. On the L&BR, the switchmen were provided with red and white flags for controlling traffic during the day and a three-coloured lamp for the hours of darkness. The distance between trains was determined by the policeman and it was left entirely to the officer's discretion as to what time interval should be allowed, leading to the inevitable possibility of mistakes.

*Railway policeman as 'switchman'*

If the early policemen were loaded with responsibilities they had little in the way of enforcement powers. The only authority the men had (I can safely say men because there were no policewomen in those days) was that conferred on them as special constables. These, if sanctioned by the local magistrate, provided for the power of arrest on company premises only. It wasn't until 1838 that railway police became an official arm of the state, when they were authorised under an Act of Parliament to keep the peace on any 'railway, canal or other public work'. For this they received five shillings a day. Each railway was required to provide officers with a uniform, great coat, hat and boots. The colours of the uniform however were not standardised and varied from company to company. In the first half of the 19$^{th}$ century, the policemen's recognised headgear was a top hat. This, not surprisingly, was impractical and within a few years was replaced by a flat cap and later the standard bobby's helmet, albeit, bearing the Company's own badge.

Railway policemen's main line of work, when they weren't acting as signalmen, was the prevention and detection of theft. Despite awesome deterrents to offenders – e.g. in 1875 a ten year old boy in Hull was sent for ten years detention on a 'training ship' for *'stealing a small quantity of cheese'* – pilfering from railways was endemic. One spectacular example was a gold robbery on the South Eastern Railway (SER) in 1854.

The SER regularly transported shipments of gold bullion between London and the Continent, a fact that became known to a well-dressed well-spoken crook called Edward Agar. Agar set his sights on relieving the SER of this heavy burden. To do so, he recruited an army of accomplices, many from the railway's own staff, with the object of attempting an audacious assault on the London to Paris train. It was a formidable task. The gold was well protected. It was carried inside locked boxes placed inside locked safes inside a locked car. The safes were also weighed three times along the route to ensure the contents weren't tampered with. Agar, however, left nothing to chance. He groomed the right people on the inside to obtain copies of all the keys he needed. He then travelled on the train between England and France several times to identify the time intervals between each stop and the various checks made along the way. On one occasion, he even sent a dummy package of lead shot on the train so he could see at first-hand how shipments of gold were processed. He finalised his plans only after making a couple of dummy runs to test the keys he had had made.

Agar waited until a large consignment of gold was on its way and then, suitably disguised and accompanied by an accomplice called Pierce, boarded the Paris train at London Bridge. The men carried with them large leather bags filled with bags of lead shot. These were deposited in the guard's van next to where the safes were located and, once the train was moving, one of Agar's inside men opened the door to the guard's van. Using the already tested keys, Agar and his accomplice quickly exchanged the contents of the safes for the lead. This they completed long before the train reached Folkestone on the first stage of its journey. The two men nevertheless stayed with the train to Dover before collecting their gold filled leather bags from the guard's van. They even had the cheek to hire a porter to carry the bags out of the station for them. Meanwhile, the lead filled safes passed safely through the scheduled check weigh-ins and it wasn't until the train reached its destination that the theft was discovered.

It was nearly the perfect robbery. Agar's undoing only came through his regular philandering. One of his many girlfriends had been present when the gang melted down the gold prior to divvying it up. A few months later she thought, wrongly as it turned out, she was going to be dumped in favour of another woman, and went and spilled the beans to the police. The principals involved were all tried, found guilty and transported to Australia to serve various terms of imprisonment.

As with insurance claims today, criminals sought to make money by feigning injuries allegedly caused by accidents on the trains on which they were travelling. One such claimant, after getting a relatively minor bump, sued the railway company for damages then pathetically hobbled into court swathed head to foot in bandages. With many of the jury in tears he was awarded £1000 in compensation for his terrible injuries. The Railway Police however were unconvinced and followed the man to Switzerland where, on arrival, he swapped the bandages for skis and enjoyed an exciting and athletic holiday at the company's expense. Unfortunately, on return to England, the police were waiting for him and he was arrested and sent to prison.

Fraud was a constant feature of railway police investigations, not least those perpetrated by the staff of the companies themselves. One notable example involved the Chief Registrar of the Great Northern Railway (GNR), Leopold Redpath whose light fingered activities became the yardstick for those that followed. Over a ten year period, Redpath got his hands on hundreds of thousands of pounds of his employer's hard-earned. He did this by systematically fiddling the company's books from his elevated position of authority within the company. Whenever the auditors' suspicions were aroused, he would boldly launch his own enquiries and after a 'thorough' investigation confidently assure them that their concerns were unfounded. Despite getting an annual salary of just £250, in just a few years he was the owner of a town mansion, a smaller house in fashionable Regents Park (each with their retinue of paid servants) and a large 'country' house in Weybridge. He even had his own personal valet to accompany him on his many regular trips to the continent. He gained a reputation as an art expert and built up a gallery of expensive paintings, one of which he purchased by outcompeting the other main bidder, Napoleon III, at an auction. In all his dealings it must be conceded he was never less than philanthropic with GNR's money. He gave large sums to charity for which he was eventually rewarded by being made the Honorary Governor of Christ's Hospital.

So how did he get away with it? Well, when challenged over his extravagant lifestyle, he claimed he was a successful investor in stocks and shares. This was true, since he printed and marketed his own fraudulent company shares which were sold to unwary punters, a lucrative practice which netted him more than £200,000 in 1856 alone. Sadly, he finally came unstuck when someone queried the huge sums being paid in shareholder bonuses compared with the actual number of shareholders on the Company's books. Redpath, however, had by then established a network of informants and was tipped off that the police were on to him. He immediately made his escape to France but soon discovered he didn't like living abroad much and didn't like foreigners at all. He therefore came back and voluntarily gave himself up. His dislike of all things foreign would prove ironic as it turned out as the penalty for his nefarious activities was transportation for life.

The Railway Police inevitably also had their share of grizzly crimes to investigate. A number of gruesome murders took place in railway tunnels in the days before automatic lighting on trains. The line between London and Brighton was especially favoured, if that's the word, in this context. This may be because it has conveniently placed long tunnels at both ends of the line. The mile long tunnel at Merstham, for example, was the scene of one particularly unpleasant incident. Around midnight on the 24th September 1905 the body of a young woman was discovered by a gang relining the tunnel brickwork.

She was at first thought to have accidentally fallen from a train but closer inspection revealed she had been deliberately suffocated with her own silk scarf which had been forced down her throat. There was no form of identification on the body so it was months before the woman was formally identified and an appeal for information launched. Witnesses then came forward and reported they had seen her in the company of a man at Victoria Station. The pair had boarded a train bound for Brighton and were later seen, by the signalman at Purley Oaks, brawling in their compartment. A man answering the same description was also seen leaving the train at the first stop beyond the Merstham tunnel, Redhill. The family of the woman was interviewed at length to establish her whereabouts on the fateful day but investigations were hampered by the non-cooperation of her brother, a fact the police attributed at the time as a worthy attempt at protecting his sister's reputation. With little further evidence to go on the matter might have ended there but for a gruesome multiple murder seven years later. The murderer, Robert Money, killed his own two children and attempted to murder his wife before setting fire to his home and committing suicide. Murray, apparently, had led a double life, having made a bigamous marriage with his wife's sister who, unlikely as it sounds, thought him to be an American with the name Charles Mackie. It transpired that Robert Money (alias Charles Mackie) was the unhelpful brother of the woman found in the Merstham tunnel, Mary Money.

Everyone is familiar with the Great Train Robbery of 1963 when 2.6 million pounds in used bank notes was stolen from a mail van by a 17 strong gang who hijacked the mail train by fixing signals at Ledburn in Buckinghamshire. This crime was so notorious that a special crime squad was set up to deal with it led by Scotland Yard. However, the Railway Police were involved in the investigation of mail train robberies ever since the appearance of mail trains. J.R. Whitbread recounts one particularly fascinating tale from those early years. This involved the arguably more audacious theft of mail from not one but two trains on the same day by the same people.

The railway targeted by the criminals was the Great Western (GWR). Late one evening in 1849, two men (Harry Poole and Edward Nightingale), caught the night mail from Plymouth, seating themselves in the carriage behind the Post Office 'tender', a locked unit connected only to the Post Office mail sorting van. In the mail van bags were sorted then transferred to the tender and locked away. The robbers waited until they were happy the sorting was completed then, as the train left Bridgewater Station, climbed out on to the carriage running board and worked their way along its side until they reached the door of the tender. Between them, they broke into the van and systematically set about opening the mail bags. With insider knowledge of what to look for, they knew exactly what each bag contained and what they were after were banker's parcels and registered letters. These they segregated and, when the train slowed, on approach to Bristol, they dropped their ill-gotten gains on to the track and made their escape. This was not the end of their adventure however. At 1am they turned up again at Bristol Station and boarded a mail train travelling in the other direction to London from Plymouth where they repeated the exercise. However, this time they had attracted the attention of a fellow passenger named Lee. He was standing on an otherwise empty station platform at Bristol and noted that the two men always kept in the shadows and conversed in whispers. His suspicions were aroused further when the men waited until the very last second before boarding the train, then rejected the compartment offered to them by a porter, for the sole reason it contained another passenger. When the train reached Bridgewater, Lee went to the station manager and voiced his concerns. The police boarded the train and caught Poole and Nightingale red handed, not having had time to dispose of their swag.

Most police investigations are not as exciting. Often they involve just the avoidance of paying fares. Nevertheless, there are some interesting examples quoted in Whitbread's book. One involved a street musician who was caught transporting his small daughter inside the bag in which he carried his harp. There was also a man who rode all the way from Holyhead to Chester hanging desperately from a brake rod underneath a carriage. For this he received a hefty fine. Another freeloader, using the same technique, was luckier in this regard. He was only fined 2/6 (12.5p) for travelling in the same perilous manner all the way from Euston to Rugby. It seems, in this instance, the relatively small fine imposed was because the magistrate was impressed by the bravery shown by the man concerned. Penalties for offences in those days were arbitrary at best. In 1875 two children were given 12 strokes of the birch for throwing stones at a passing locomotive.

*Victorian railway policeman*

In 1949, following nationalisation of the railways, the separate railway police forces combined to form the British Transport Police. Nevertheless their story doesn't end there. They have been the authority and reassuring presence in the aftermath of some of the most infamous events in recent history including the Kings Cross Underground fire in 1987 and the '7/7' bombings of 2005. These days, however, their role in combating crime overlaps to a much larger extent with that of the regular police, especially when it comes to such matters as terrorist activity and international criminality, when the role of the railway in a crime may be only incidental to a wider network of skulduggery.

So what do we know about John Robert Whitbread himself? The answer is very little. Neither the British Transport Police History Group nor the Transport Police themselves could provide much information on him[24]. However, it is known that, following the publication of Whitbread's book, a subscription was raised to fund an award, in his name, for 'exceptional behaviour' by railway policemen. The Whitbread Shield, in the fifteen years or so of its existence, was awarded to only a select number of officers whose actions went far beyond the normal call of duty.

I started this article with one of Whitbread's stories about unhinged users of our railways and I'll finish with another. For a few months, in 1959, British Railways (BR) were plagued by a series of weird faults on their signalling systems near Hull. There was, for instance, inexplicable mutilation of speed restriction boards, damage to telephone installations and, most worryingly of all, interference with points and signals, including the removal of the batteries that energise track circuit indicators. This left signalmen with no knowledge of where trains were located on the line. On top of this potentially deadly vandalism BR also discovered the top two feet of fences beside their railway lines was systematically being removed. Since the attacks were indiscriminate and random months went by without the police able to identify the culprit. It was a police dog that eventually came to the rescue. Following a tip-off from a landowner that a man was seen chopping down trees on a plantation next to a railway tunnel, the dog successfully pursued and caught the miscreant. Once in custody, the man happily confessed to all the crimes. He was, he said, acting in the interest of his fellow travellers. Signals, he believed, were distracting to train drivers and needed to be removed as a safety precaution. When asked about the shortening of fences he told the police that this simple procedure would enable trains to go faster although the physics of how this worked was never explained. Not surprisingly the man was committed for psychiatric treatment.

---

24 In fact it wasn't until 2014 that he was identified.

**Notes to Chapter 20**

I am indebted to Bill Rogerson MBE of the British Transport Police History Group for his assistance in providing background to the above. I would also like to thank Malcolm Clegg of that group for letting me use the picture of John Whitbread – as far as I am aware the only photograph of the man in the public domain.

# Postscript

When I compiled the magazine articles that made up 'Those Railway People', I had no idea that anyone other than family and friends would want to read the resulting book. In consequence, I didn't take as much care over it as I should have. In particular, I had the naïve belief that a compendium of previously published magazine articles would be able to stand alone without major revision. Big Mistake.

The main problem was repetition. My stories were written over a period of twenty years, and reproduced in a variety of publications. It didn't matter at the time, therefore, that some of the same ground would be covered from article to article. Put together in book form, it mattered a great deal. When I recently re-read the book I was embarrassed by the number of times the same events were reported, albeit in a different context. The first thing I needed, therefore, was to excise the sections which told the same story; in one instance this involved removal of a complete chapter. I would like to say there is no repetition in 'Railway People', however this is not so. With the best will in the world, when writing different stories about the same people and places you need to repeat sufficient material to make each chapter self-contained, without the necessity for the reader constantly referring back to earlier chapters. To cut out all repetition would have made the book harder to follow. I hope, therefore, I have achieved the best balance possible under the circumstances.

Revising 'Those Railway People' did give me the opportunity to bring it up to date and correct mistakes. I like to think (he said modestly) I know more now about the early days of railways than I did twenty years ago. When, for example, I originally wrote about John Wesley Hackworth's trip to Russia, I had no idea that Thomas Hackworth was involved in the locomotive's manufacture. Indeed I thought, incorrectly, that Tom Hackworth had already left his brother's works and moved away to Stockton by the time the engine was built. These and other errors and indiscretions have, hopefully, been corrected.

Finally, I added a few of the pieces written since 'Those Railway People' was published in order to bring the collection up to date while, at the same time, changing a few of the pictures. For those who purchased the original book, I hope you like what I have done with it. And for those who are reading these stories for the first time, I hope you find them entertaining.

George Smith
(September 2016)

# Just a selection of the bibliography used in the research for this book

1. Gerin W. 'Branwell Bronte – A Biography'
2. Gaskell E. 'The Life of Charlotte Bronte'
3. Bentley P. 'The Brontes and Their World'
4. Tomlinson W.W. 'North Eastern Railway'
5. The Minutes of the Finance Committee of the Manchester and Leeds Railway 1840 to 1845
6. Rennie J. 'Autobiography of Sir John Rennie FRS' (London:1875)
7. Skeat W.O. 'George Stephenson – The Engineer and His Letters' (Inst. of Mechanical Engineers:1973)
8. Dendy Marshall C. 'History of the Southern Railway'(Ian Allen Ltd.:1963)
9. Oppitz L. 'Surrey Railways Remembered'
10. Lowe J.W. 'British Steam Locomotives'(Goose and Son:1975)
11. Stephenson R.'Report on the Atmospheric Railway System'
12. Hadfield C. 'Atmospheric Railways – A Victorian Venture in Silent Speed'
13. Minutes and reports of the London and Croydon Railway Company
14. Griffiths E. 'The Selsey Tramways' (E.C.Griffith:1974)
15. Bathurst D. 'The Selsey Tram' (Phillimore:1992)
16. Cooksey L.A. 'The Selsey Tramway' (Vols. 1 & 2)'
17. Minutes of Hartlepool Dock & Railway Company
18. Sharp C. 'History of Hartlepool' (John Procter:1851)
19. Thornley Coal Company records
20. Fay C.R. 'Huskisson and His Age'
21. Garfield S. 'The Last Journey of William Huskisson' (Faber and Faber:2002)
22. Davies H. 'George Stephenson' (Quartet Books Ltd: 1977)
23. The papers of Leveson-Gower, 1st Earl of Granville
24. Rolt L.T.C. 'George and Robert Stephenson – the Railway Revolution'
25. Hogg P.L. 'Richies – A History of Thomas Richardson and Sons and Richardson Westgarth 1832 to 1994' (Hartlepool Borough Council:1994)
26. Castle Eden Parish Register 1750 to 1820
27. General Minutes of the York, Newcastle and Berwick Railway
28. Stretton C.E. The Development of the Locomotive – A Popular History 1803 to 1896'(Crosby, Lockwood and Son: 1897)

29. Vol. 5A of the British Locomotive Catalogue 1825 to 1926
30. Sidney S. 'Rides on Railways ( with a new introduction by Barrie Trinder)' (Phillimore and Co: 1973)
31. Sidney S. 'Rough Notes on a Ride over the Track of the Manchester, Sheffield, Lincolnshire and Other Railways'
32. 'Graces Guide – The Best of British Engineering 1750-1960s'
33. Smiles S. 'Life of George Stephenson' (London:1857)
34. Bealby R. 'Stockton-on-Tees Local History Journal no 4 Spring 2003'
35. Kirby M.W. 'The Origins of Railway Enterprise'
36. Minutes of Hartlepool Dock and Railway Company
37. Rolt L.T.C.'Isombard Kingdom Brunel'(Longman Group Ltd.:1957)
38. Nock O.S. 'The Great Western Railway in the 19th Century'(London:1962)
39. Brooker F. 'The Great Western Railway – A New History' (David and Charles:1977)
40. Vaughan A. 'The Intemperate Engineer – Isombard Kingdom Brunel in His Own Words'
41. Hale M. 'The Oxford Worcester and Wolverhampton Railway Through the Black Country'
42. Boynton J. 'The Oxford Worcester and Wolverhampton Railway'
43. Jenkins S.C., Quale H.I. 'The Oxford, Worcester and Wolverhampton Railway'
44. Coleman T. 'The Railway Navvies'
45. Macdermot E.T. 'History of the Great Western Railway'
46. The Committee Minutes of the Oxford Worcester and Wolverhampton Railway
47. 'Beeching: Champion of the Railway?' by R.H.N.Hardy
48. Henshaw D. 'The Great Railway Conspiracy' ((Leading Edge Press:1991)
49. Wolmar C. 'Fire and Steam' (Atlantic:2007)
50. Marr A.'A History of Modern Britain'
51. British Railways Board 'The Reshaping of British Railways'
52. Hansard: 10th March 1960
53. Letter of Minister of Transport Ernest Marples to Dr.Richard Beeching dated 6th April 1960
54. Beeching R. 'Interpretation of the Terms of Reference of the Special Advisory Group' (3rd May 1960)
55. Special Advisory Group on the British Transport Commission 'Recommendations to The Ministry of Transport and British Transport Commission' (July to September 1960)'
56. Swinger P. 'The Power of the B17s and B2s' (Oxford Publishing Co:1988)
57. Whitbread J.R. 'The Transport Policeman' (G.Harrop and Co: 1961)

58. Young R. 'Timothy Hackworth and the Locomotive' (The Book Guild:1923)
59. Burton A. 'Richard Trevithick – Giant of Steam' (Aurum Press:2000)
60. Holmes P.J. 'Stockton and Darlington Railway' (First Avenue Publishing)

In addition to above, in respect of the 'Battle of Mickleton Tunnel', the Illustrated London news dated 26/7/1851 and Berrows Worcester Journal dated 24/7/1851 were useful sources.

# Acknowledgements (and apologies)

It would be an impossible task to acknowledge all the people who contributed in some way to the above over many years. However, there are some candidates I must single out for special thanks, namely:

- Staff of the National Railway Museum at both York and Shildon
- The National Archive at Kew
- Beamish Industrial Museum
- The British Library
- 'Head of Steam' at Darlington, North Road
- Staff at Durham County Council Archive

And last but not least my wife Maggie who had to put up with it all.

Finally I apologise to those readers expecting a comprehensive list of reference sources. Where I did keep complete records I have updated the original list (which appeared in 'Those Railway People'). Where I didn't, I've left the references as they were. To be honest I just couldn't be bothered trawling through all the incomplete references to identify sources and dates. Life's too short.

Most of the photographs in the book are my own or were reproduced from journals and books dating back in some instances more than a hundred years. I would like to thank again Denbighshire CC for letting me reproduce the picture of the Llangollen rail disaster, Chichester Museum for the Selsey Tram picture, John Crocker for the picture of the ex GWR Castle at Mickleton, Malcolm Clegg of the British Transport Police History Group for the picture of John Whitbread and various fans and officials of football league clubs for the 'football' B2 pictures.

If I have inadvertently missed out any other helpful 'railway people', I once again apologise.